AF710648

INVENTION WORKS

VOLUME 5

ETHICAL TECH

Technology Innovations and the Environmental Ethic

INVENTION WORKS

VOLUME 5

ETHICAL TECH

Technology Innovations
and the Environmental Ethic

RICHARD CRANGLE

President and Founder

CRANTEC RESEARCH

BIOMEDICAL TEST TECHNOLOGIES

CLINICAL INNOVATIONS OF SOUTH DAKOTA

Adjunct Professor

UNIVERSITY OF UTAH

SALT LAKE COMMUNITY COLLEGE

WESTMINSTER COLLEGE

Cranstone Publishing House
Salt Lake City, Utah 84121

Invention Works Series

Invention Works: Volume 1
Success
Develop, Protect and Make Money with Your Invention
ISBN: 0-9664835-1-0

Invention Works: Volume 2
Piracy
Understand Piracy and Protect Your Patents, Copyrights, Trademarks and Trade Secrets
ISBN: 0-9664835-2-9

Invention Works: Volume 3
Healing Tools
The Development of Medical Theories, Innovations and The Quest for Health
ISBN: 0-9664835-3-7

Invention Works: Volume 4
Disruptive Devices
The Risks and Rewards of Highly Innovative Concepts and Technologies
ISBN: 0-9664835-4-5

Invention Works: Volume 5
Ethical Tech
Technology Innovations and the Environmental Ethic
ISBN: 0-9664835-5-3

All rights reserved. This book is protected by copyright. No part of this book may be used for reproduction, stored in retrieval systems, or transmitted in any form or by any means, electronic, mechanical, photocopied, microfilmed, recorded, or otherwise, without written permission from the publisher or author.

Copyright © 2008
Richard Crangle
Salt Lake City, Utah

Cranstone Publishing House
Salt Lake City, Utah 84121

Library of Congress Number:

ISBN 0-9664835-5-3

Printed in the United States of America

CONTENTS

FORWARD

The cover graphics of ***Volume 5*** include a depiction of Diogenes, a Greek thinker of the 4th century BCE who went about Athens with a lantern seeking to find ethical persons. Historians recount his meeting with Alexander the Great who came to visit Diogenes at his Athenian home that was in a large barrel. Unimpressed with Alexander's power, Diogenes requested that Alexander not block the sunlight by which Diogenes was reading. The disciples of Diogenes became known as "cynics" derived from *cynos* the Greek word for dog and founded the counter-cultural school of the Cynics. Their uncompromising loyalty and fidelity to the teachings of their master who disdained the trappings of political, economic and religious authority caused them to live in "dog houses."

In the post-modern world dominated by the power of technology, money, military and corporations, the search for ethical persons continues despite a vast increase in population. Although 23 centuries separate Diogenes from the inhabitants of the 21st century, many of the challenges remain the same. One continuing challenge is to establish a credible ethical system to guide moral decisions and behaviors that includes preservation of the environment in which all inhabitants live, move and have their being. The primary task of environmental ethics is to help sustain eco-systems that support bio-diversity and provide a place for multiple forms of life to make a living. This requires more of us to carry our lanterns, speak authentic truths to power and preserve a breathable atmosphere that also allows us to see the light.

MEANINGS & TERMS OF ART

Certain key words or phrases appear in the following chapters that are Greek or Latin and are retained in their original language. They continue to be stated in the Greek or Latin in various types of literature since they have taken on fixed denotations and connotations.

Other commonly used words such as "stewardship" have been used broadly and in many contexts. This usage pattern often obscures the etymological origin of the words which conveys a message and meaning distinct from the acquired connotations.

Some words such as *aretean* are used as terms of art that have developed to better express certain concepts, categories or patterns of analysis. Aretean, for example, is derived from the Greek word *arete* that means excellence and, as a term of art, is used to identify thoughts and actions that aspire to the highest standards or ideals of excellence. Similarly, *utopian* is a Greek word that literally means either "a good place" or "nowhere" in that it refers to an ideal place that exists only as an *aretean* concept but not in the "real world" of percepts.

"Radical" used as a noun or an adjective has taken on connotations that suggest a myopic view and has become synonymous with fanatical. However, "radical" is derived from the Latin word *radix* meaning "root" wherein the denotation has referred to those seeking the sources of nourishment for an organic entity. A combination of its denotation and connotation in contemporary expressions suggests a lack of willingness to compromise premises or tenents of belief.

Pragmaticism and pragmatism are terms derived from the Greek word *pragmatis* which denotation means "to do business" and its connotation includes political and legal functions. While pragmatic practices have been employed over many centuries, the words were specifically introduced into ethical and moral discussions as terms of art by C. S. Pierce and Williams James in the mid-19th century. Pierce used pragmaticism, a word he thought "too ugly to be stolen," while James preferred pragmatism as a guide for moral choices, some of which are "the moral equivalent of war." Despite the evocative talents of these writers, pragmatism in its various forms fails to address significant ethical issues and begs most every moral question. That an action works is a trite and empty justification for any behavior and affirms the denotation of doing business as usual, as an expedient or as a practical matter.

Central to the discussion of the environmental ethic are the terms "ethics" and "morals" that are also common currency in almost all literature that addresses standards of human behavior. Ethics is derived from the ancient Greek word *ethos* that contains the essential element of a standard that is inherent in its denotation and connotation. Its etymological Endocuropcan root conveys immutable standards not subject to social or cultural relativism that is described as changing sociological or anthropological phenomena. This element of ethics as a fixed star of meaning is found in related words such as "ethereal" or "ether" that are derived from the same root. From this realm, rational ethical theories and systems are derived and become the foundations for ethical "ought" propositions.

In contrast, the word "moral" has its etymological origins in the Etruscan-Latin word *mores* that means custom or convention within a tribe. Morals as customs are mutable in that the conventions or behaviors within tribes change over time and display disparity of cult. Moral behaviors are studied by natural and social scientists as patterns to be observed and described. Although subject to flux and the vicissitudes of historical dynamics, moral retains the elements of contextual standards in that it refers to gradient minimum and maximum limits of human behavior within a social group. For example, killing other human beings in war and despoiling the land of an enemy were both allowed as a minimal standard for citizens of Rome from 450 B.C.E.-476 C.E. and were also extolled as maximum expressions of *virtu*, the gold standard of moral behaviors in that time and place. During the *Pax Romana*, killing fellow citizens was usually redressed by the imposition of a money fine; killing one's infant child by "exposure" was morally and legally allowed under the Justinian Code provided that the parents notify three neighbors.

When these words or phrases appear in the text, there is an explanation stated in an endnote to translate or clarify the meaning. An awareness of these distinctions will be helpful to those interested in the quest for credible ethical standards that may form the foundations for applied morality. Moreover, those seeking a secure theoretical structure for ethical guidance within the realm of environmental ethics need to reflect on and discern the real origins of their *ethos*. Similarly, it behooves serious students of human behavior to understand the metes and bounds of acceptable treatment for our environment wherein our mutable lives may continue to exist or be extinguished based on our chosen values and our actions.

PREFACE

Ethical Tech is the fifth volume in the **Invention Works** series that is focused on technological innovations and their commercialization. As indicated in previous volumes, inventions emerge from within a variety of environments, from ingenious individuals or collective efforts, from contexts of urgent need and continuing resistance. Some significantly innovative technologies have been developed and disseminated rapidly such as the production of eye glasses in the 13th century, while others such as medical devices, agricultural implements, solar heating and cooking appliances have languished for centuries, despite continued need and necessity.

Still other significant innovations have been ignored, even strenuously resisted despite overwhelming evidence supporting the discovery and the dire need for a solution. Giving citrus fruit to sailors to prevent fatal scurvy was not embraced by those regulating maritime activities but was opposed for over 50 years after its benefits were proven. Inoculation for dread diseases such as small pox was resisted by both the medical profession and governmental agencies although its efficacy had been clearly demonstrated by Edward Jenner in 1796.

Paradoxically, some innovations and discoveries that are dangerous or patently harmful to health are produced and proliferate in the stream of commerce. Tobacco, for example, introduced from America into a global population in the early 17th century has caused primary and secondary diseases with outcomes as lethal as small pox or malaria epidemics. Four hundred years later, tobacco products are protected by governmental agencies established to safeguard health and they generate hundreds of billions of dollars annually in revenues. Healthcare required by tobacco-related diseases also costs hundreds of billions of dollars annually worldwide much of which cost is borne by taxpayers.

The anomalies, paradoxes and outright contradictions of developing harmful technologies have been staples in the inventory of innovations since the anatomically modern man fashioned the first stone tool to kill animals and other humans. In addition to the ongoing challenges of researching, designing a new product concept, developing it and disseminating it in the stream of commerce, serious questions regarding its position within a hierarchy of values including its affects on the environment need to be answered. Even from a strictly utilitarian standpoint, it must be asked whether or not any technology innovation is justified in terms of its consequences such that it produces more total "hedons" of satisfaction for the greatest number of people in the long run.

Regardless of the technology innovation involved, the consequences also need to be assessed by other arbiters using alternative criteria for evaluation. These arbiters include divergence among utilitarians themselves wherein ethical disputes erupt from those wanting to use set "rules" to assess short-term, intermediate and long-term consequences. Other utilitarians demand adherence to "acts" on a case-by-case basis to determine the total hedonic calculations from the consequences of certain acts.

Apart from the widely popular and enduring appeal of evaluating consequences of human behavior using a utilitarian index, there are many other traditional models that have sought to assess moral actions. A virtual cafeteria, offered over the past 24 centuries in Western thought, has competed for attention and acceptance from those seeking ethical foundations and moral guidance. Attention has been paid to the leading ethical positions and systems that have stood the test of time by being repeated and honored by committed cadres over the centuries despite sharp criticisms launched by their adversaries.

These classical conflicts in theory, application and practice are seen in contemporary debates regarding ethics in every area including environmental ethics. Most all of the questions and critiques of ethical theories, models and systems generally are also central to the debates surrounding technology and environmental issues that impinge on ethical concerns and moral behaviors. One of the recurrent disputes central to ethical issues concerning the environment involves the distinction between ethical ideals and the realities of ethical applications. Ideals defining the highest and best standards of excellence have been given varied definitional content over time as have those espousing the need to formulate nuanced public policy. Whatever the particular definitions given by each of these disputants, the unresolved tensions are generated from the refusal of the idealists to compromise their ideals with nuanced public policy formulations.

These perennial controversies are brought into discussions regarding environmental ethics and technology that show a polarity between those urging the ideals of the "wilderness concept" and their adversaries wanting modification of this ideal in the form of public policy formation such as "mixed use." In sharp contrast to the idealists' language incorporating irreducible "good", categorical imperatives, inalienable human rights and absolute pollution bans, those influencing legislation, such as corporations, politicians and the courts interpreting legislative enactments, use alternative agendas.

While these ideological and practical quarrels continue in the multiple forums of ethical debate, the environment continues to be affected by human technology and consumerism in many ways detrimental to it. Populations are exceeding the carrying capacity of the earth. Atmospheric pollution, including greenhouse and toxic gases, contribute acid rain and global warming increases, while non-renewable natural resources are being depleted. Technology innovation is failing to provide solutions to problems of its own making. Within the context of these multiple crises, ethical issues are confronted and moral behaviors are being evaluated. It remains to be seen whether a viable and useful environmental ethic can be developed that will contribute in a tangible way to the sustainability of the earth, its inhabitants and its eco-systems. Unless such an environmental ethic can be crafted immediately, critical decisions will continue to be made by default and inactivity. Crucial conversations with those who continue to allow the ongoing predation and pollution of our planet are urgently needed to confront and address these central moral issues of the 21st century.

INTRODUCTION

A didactic story used by Jean Bodin, a logician in the late Middle Ages, graphically illustrates the difficulty of decision-making. Bodin's story involves the dilemma of a donkey that is faced with choosing between two desirable piles of grain. Despite prolonged analysis, contemplation and discernment, the poor creature was unable to decide between the two thereby becoming impaled on the horns of the dilemma. The result was that Bodin's donkey starved to death.

As with many medieval tales and their extended versions in morality plays during the Middle Ages, the stories speak about the human condition and its many predicaments. These predicaments involve decisions about behaviors, plans of action and life strategies that take place across the spectrum of ordinary lives that sometimes may demand extraordinary decisions. Some of these decisions are seemingly simple, while others may be forced upon us with little or fragmentary information upon which to base a decision. Still others may be created by the various *necessitas vitae presentis*, the necessities of one's life in the present tense as the medieval writers have stated. These decisions are often both urgent and important and they may come with a wide range of relevant informational content, some which may be correct while other content may be false. Within the context of these competing issues, interests and concerns, decisions including ethical choices must be made, despite multiple elements of indeterminacy along with cognitive and emotional dissonance.

Multiple voices from the past and present offer or demand control of our future choices whether public or personal, whether self-interested, self-considering or self-serving. These voices include seductive songs from Sirens not unlike those that caused Odysseus to be lashed to the mast lest his ship be impaled on the rocky horns of Scylla and Charybdis. The Sirens' chant was an ancient version of Friedrich Neitzsche's "lovely lie of sound" that promised Odysseus perpetual glories of the past rather than future rewards or endless pleasures. Many of these Sirens' songs are expressed in a wide spectrum of ethical advice over the centuries amplified by centers of authority that define what is orthodox. Aristotelian writings from the 4th century B.C.E., for example, that taught an eudaemonistic system of ethical decision-making commanded pride of place among throngs of competing advice-givers. When made orthodox through the adaptations of Thomas Aquinas in the 13th century, 50 years after the Condemnation of 1277, the status of *eudaemonism* was greatly augmented by "tuning the pulpit," the method of media control of the dominant Church for subsequent centuries.

Other teachings emerging over the centuries encouraged, directed and coerced both willing and unwilling audiences to direct their lives and behaviors by making certain ethical choices with the expectation of happiness. Those choices were encouraged or demanded and the promise was that some kind of eudaemonism or happiness would result. However, the term, *eudaemonia* used by Aristotle actually means controlling one's demons, although it has been erroneously translated for centuries as "happiness." *Happe* is a French word that means a lucky chance occurrence that certainly leaves serious ethical decision-making at the mercy of the capricious fates. There have also been competing voices as to the ways and means of getting to this elusive blue bird of happiness that range across a spectrum of deontology, act and rule utilitarianism, social contract theory, hedonism, situation ethics, moral points of view, natural law, theological, moral sense, theories of justice, inherent principles and versions of all of these. These systems and their permutations of metaethics and applied ethics claim to produce correct ethical choices and moral behavior by following their norms.

These promises were based on normative formularies that were said to produce happiness in the past. By following these normative teachings that also defined certain values, motives and behaviors, the ethical person would share in a state of being deemed to be the ethical 'gold standard' of the past. In this way, scripts for each person were written that defined roles to be played in the theatre of life. For Neitzsche and other existential ethicists, the authentic moral person's primary task is to "overcome this [one's own] linguistically generated self."

Despite the broad range of ethical foundations on which to base our choices, the outcomes generated from these various theories and applications are remarkably similar. Victimization is a factual issue not created by ethical or legalistic theory. Once a victim is clearly defined, all the traditional ethical systems produce prescriptive conclusions to spare the victim from harm. Followers of most any one of the traditional ethical schools of thought would come to the same conclusion once a victim is identified, the factual evidence known and understood. In the so-called "hard cases" of killing human beings whether in war, abortion of fetuses, euthanasia or for love and money, once the facts are agreed upon, almost all ethical theories point in the same direction. If they do not align, the ethical theory or application is presumptively fatally flawed; its premises misunderstood; a correct ethical point of view is not taken; or assertions of legal or political interests are mistaken for authentic ethical considerations.

Ethical decision-making usually does not lack common ethical prescriptions, norms or value-laden language; the primary problem of actually choosing the right course of action and not acting on it is primarily the failure to control one's demons. Aristotle's luminous insight unfortunately has been obscured by translating *eudaemonia* as happiness and sending countless generations of seekers on a futile quest to find it. Aristotle's definition of virtue is a state or habit of character residing in a mean between excess and defect and relative to ourself or as a person of wise counsel would state it. However, virtue is not cultivated to bring about happiness and may well cause unhappiness at least for a while. Virtue is developed, often with great discipline and difficulty, to respond to our personal demons and to help control them. In fact, sturdy and purposeful virtue is necessary to protect oneself and others from the ravages of at least the seven deadly sins and capital vices that spawn other destructive behaviors. Lucky chance occurrences may or may not happen to bring about *happe* and are essentially irrelevant in seeking the "golden mean" in our lives that is needed for both the happy and unhappy life.

A virtue-based ethics articulated by Aristotle has been revisited by contemporary scholars of ethics and those in the area of environmental ethics. This approach holds significant promise for the development of a cogent environmental ethic and moral behaviors with respect to responsible and sustainable ways of living. With the frailties, short-comings and failures of alternative ethical theories and moral guidelines, cultivation of virtues that includes relevant indices appropriate for environmental health would place specific responsibility and accountability on an individual as a primary moral agent. An environmental virtue such as personal frugality would become a habit and a state of character by repetition wherein the environmentally virtuous person would become an ethical person identified by personal action consistent with that virtue.

Proliferation of such virtues could be accomplished by various means including educational organizations and multiple media outlets that have heretofore promoted models of conspicuous consumption, gluttony and mindless consumerism all of which are environmentally destructive. Launched from a multiplicity of platforms, models of environmental virtue would shift behavioral paradigms and proliferate as "memes," the social equivalent of genes that impart their designs or phenotypes on forms and ways of life. It remains to be seen whether a 'culture of environmental virtues' could be cultivated wherein organizations themselves would function according to an adopted list of virtues.[1]

Regardless of the individual or social belief system that corresponds to one of the many ethical systems selected, the person seeking to become ethical first needs to get clear about ends, means and meaning. Once the ethical choices are reviewed, an array of means need also to be considered with the desired end selected that is to be achieved. In this process of discerning appropriate means, an ethical system may be adopted and used to justify the choice of a desired outcome before or after the fact. As a form of justification, an ethical system or theory does little to guide ethical decision-making but serves as a means to provide validations for a decision already made.

Rather than searching for an ethical theory to make ethical choices, it is possible to locate a variety of systems to justify the choice of any behavior driven by other dynamics, including our unrestrained demons that incline us to denial, self-deception and delusions.[2] Acceptance of the basic facts regarding environmental degradation, for example, requires individual and collective reality therapy for those in the grip of consumerism that systematically promotes continuing denial, self-deception and delusional behavior. In fact, continuing to destroy the environment by compulsive purchase of vast inventories of fungible merchandise is a fairly accurate description of madness that expects different results from the same course of action.

The search for ethical theories as systems to generate specific ethical decisions has resulted in a multi-generational quest to generate the illusion of certainty from ethical theories. Ethical systems that purport to generate right ethical choices upon weighing values appeal to those who demand method and uniformity of decision-making outcomes. Such systems have been constructed on various premises ranging across the spectrum of religious and non-religious based theories. Some theologically based traditions include the Judeo-Christian methods of proliferating ethical directives in the form of summarized commandments which numbers have been reduced from the Talmudic 613 Commandments to ten in the Christian version. The summary teachings of the founder of Christianity have further reduced the Ten

Commandments to two: first, "Love God" and second, "Love your neighbor as yourself." Four hundred years later, St. Augustine condenses all of Christian moral theology into seven words: "Love God, then do what you will."[3]

A representative example of a non-religious ethical system is utilitarianism whether in its "rule" or "act" form. In the original formulation by Jeremy Bentham, value weights could be transmogrified or "morphed" into hedonic units that could then be subjected to a "hedonic calculus," then counted and weighed. The proper working of the calculus as the operating platform of the system would then generate the correct answer to any ethical question at hand. The decision-maker would be spared deliberative anguish or moral ambiguity as the quasi-mechanical process grinds out the proper course of action in each situation. Only an inept operator of such a system or a computer malfunction would cause one to become stranded between two piles of grain.

Despite the allure of these kinds of systems, the Platonic "mere appearance" of promising ethical rectitude disguises the complex and subtle ways of forging an authentic moral decision. Plato's illustrious student, Aristotle similarly warned against expecting exact or mechanically uniform results from moral judgements and counseled reasonable expectations because of intellectual limitations, natural irascible and concupiscent appetites, along with the bias of nurture within societies. Even at best, there can be disagreement since reasonable persons disagree and some moral judgements may only ameliorate evil rather than prevent wrongdoing altogether. Both recognized the premises chosen by an individual moral agent control the judgement just as the conclusion of a syllogism logically follows its major and minor premises. If a hedonist, for instance, were to enter personal pleasure as having the greatest hedonic weight, the greatest good for the greatest number in the short or long run could easily be eclipsed by totally self-serving ethical decisions. Such an outcome could be generated in every case, thereby rendering the system ethically useless and meaningless. That the qualitative values are arbitrarily given certain quantitative value in order to add, subtract, multiply or divide discrete units does not provide reliable indices or criteria to make moral decisions.

Similarly, any attempt to apply covert or overt quantification methods to other ethical theories, including social contract theories or theories of justice, results in the subtle alchemy of transforming qualitative values into quantitative units. It is admittedly appealing to revisit the quixotic illusions of the medieval alchemists who sought to transform base metals into gold. However, predicating the bonds and covenants of a social contract on this slight of hand only produces an indefensible contract, void for vagueness and unsuited to form either the social adhesions of civility or the necessary and sufficient cohesions of a social or ethical meeting of the collective minds and actions. Unfortunately, theories and principles of justice, benevolence, non-malfeasance or autonomy are not reinforced by efforts to quantify or give weights and measures to their underlying premises.[4] Even if John Rawls' "veil of ignorance" could be sewn of heavy enough cloth to actually shroud the positions, behaviors and agendas of the ethical parties of interest, the outcomes of ethical decision-making still lack predictive power to provide reliability or integrity for the prescriptions urging certain moral actions.

By using numbers, equations, commandments, principles or rules behind the "veil" with one's own personal demons unchecked or others similarly situated still in the grip of denial, self-deception or illusions, an artifice is constructed that is contrived and will be shown to be unsound in theory and unworkable in its applications. Specifically, the threshold issues concerning environmental ethics and technologies involve overcoming individual and collective denial, self-deception and delusional thinking regarding the unvarnished facts and values relevant to the many problems of the environmental ethic. It needs to be recognized that human behaviors on a multiplicity of levels directly contribute to global climate change, toxic pollution and the destruction of natural resources including eco-systems. The efficacy of any environmental ethic to ameliorate environmental degradation is initially dependent upon that level of intellectual and moral integrity. Then it remains to be seen whether individual persons and collective groups are willing to become disillusioned enough by ridding themselves of dysfunctional illusions and to take decisive moral action at the level of their convictions. If not, there is a high statistical probability that there may not be two piles of edible grain harvested from our tragically destroyed commons between which many of us could become stranded and starve to death.

Chapter One

Ethical Systems & Their Environmental Implications

The impasse created by conflicting values in ethical deliberations is more apparent and amplified when specific areas of applied ethics are analyzed. Issues involving environmental ethics, for example, which demand reliable, credible and integrated predictive capabilities, are left in the lurch with the offerings of traditional ethical systems as their arbiters. Without addressing and resolving the obstructionism and obscurantism of denial and self-deception at a fundamental level, ethical discourse and policy formulations offer little guidance and, as such, are without substance or merit. One such fundamental and underlying premise is that of anthrocentrism that exclusively favors human interests over those of other creatures and the environment itself, and also distorts dialog and decisions regarding the environment at nearly every level.

Anthrocentrism influences a wide range of ethical arguments and policies along with individual or collective behaviors; it also has been a point of view to favor human benefits. As a foundational major premise, anthrocentrism is often assumed *a priori,* not unlike Kantian space and time, such that it is usually mixed in with other premises to become a dissolved salt in the water that becomes the medium for evaluating ethical behavior. Individual and selected groups also participate in the assumptions implicit in anthrocentrism by asserting rights, prerogatives and prescriptions that give them ethical sanction for behaviors that are often self-serving, but are detrimental to other forms of life.

The implications of these issues are the subject for an increasing volume of literature that has been escalating in quantity and quality since Peter Singer's seminal book, *Animal Liberation* and his widely cited article "Famine, Affluence and Morality," Philosophy and Public Affairs J., no. 3 (1972). Even when accepting anthrocentric major and minor premises to give pride of place to human interests, fundamental ethical questions affecting the environment still remain and are often begged. Rather than providing a clear and consistent ethical compass, anthrocentrism itself fails to answer general or specific ethical questions. The interests and aggressions of individual humans and the collective expressions of these predispositions generate ethical conflicts with profound environmental consequence without even overcoming the human/non-human polarities.

Peter Singer, for example, sets forth cogent arguments to support a claim that affluent people have an ethical obligation to give food to the poor and hungry. However, critiques of such an ethical prescription include those of John Rodman and Garrett Hardin who question the practical consequences of this seemingly benevolent form of anthrocentrism.[5] For both of these critics, giving food to the malnourished and hungry would increase the demand and supply sides of food production with deleterious effects on the environment. One result would be dramatic human over-population with damaging effects on impoverished regions and indirect damaging results on the affluent regions due to immigration by the poor.

Hardin argues forcefully that the operation of self-interest within an anthrocentric framework has additional damaging consequences. In fact, he emphasizes that a general ethical prescription or a public policy that requires sharing food with hungry people is "fundamentally wrong." It is wrong, according to Hardin because it leads to the tragedy of the commons.[6] The commons refers to any environment in which certain areas such as pastureland are used by multiple persons to benefit those sharing this area. A critical premise operating in shared uses of the commons is that each commons has a limited "carrying capacity" beyond which the marginal utility decreases and that particular commons deteriorates. When used within the parameters of its carrying capacity, mutual benefit is realized by all those participating.

According to Hardin, when the supposition of self-interest is interjected into the context of a shared environment, the carrying capacity is exceeded to the detriment of the commons and those using the commons.

However, when considered on a rational basis of enlightened self-interest, each one using the commons may introduce just one extra user of the commons for individual gain but only share a small fraction of the overall costs. The tragedy results from the cumulative effects of others using the commons in the same way to further their own self-interest even though the results are ruinous to both the users of the commons and the commons itself.

The dynamics of this model, proposed by Hardin and reinforced by Rodman and others including those in the Green Revolution, are relevant to a wide range of human behaviors that have a direct and indirect affect on people, animals, places and things. Population growth as a key issue with far-reaching ramifications is seen on this view as directly causing the tragedy of the commons in most every environmental venue. Excessive grazing, over-fishing, clear-cutting of forests, depletion of non-renewable resources such as coal and oil, along with pollution of air, water and land are variations of the tragedy of the commons. Specifically, over-population can be used as an explanatory and predictive model to show how self-interest is translated into a right of couples to produce an extra child for perceived personal gain with the cost distributed over those using the commons of the nurturing and educating environment. It is clear to those employing this model that over-population that exceeds the carrying capacity of the commons can easily be predicted to occur within a measurable timeframe.

When reproduction trends are compiled and analyzed, the predictive and explanatory power of this model becomes readily apparent. Quantification of worldwide reproductive behaviors by 1970 indicated that human population in wealthy nations was doubling within 87 years while it was doubling every 35 years on average in non-wealthy nations. However, within the same time-frame, the populations of certain countries including Columbia, Venezuela, Ecuador, Morocco, Thailand, Pakistan and the Philippines are doubling human population every 21 years and populations in Mexico were growing over four times more rapidly than in wealthy nations. Whether leading to intended or unintended consequences, population growth patterns and the environmental impact of doubling populations, along with business exploitation of depleted resources, unenlightened self-interest is a central component of these dynamics.

Self-interest has been an enduring element in human thought and behavior along with its prominence as a human and civil right increasingly asserted in arbitration and litigation arenas. However venerable self-interest may be, its unenlightened forms have wreaked havoc especially when unchecked by effective countervailing forces. As a principle of human behavior, self-interest has manifested itself as a primary personal and corporate demon to be tempered through Aristotelian habits of virtue or governmental force. Without these checks, the balance of *eudaemonia* remains unreachable and various tragedies of the commons appear to be the logical and experiential outcomes.

As a principle of collective behavior, self-interest has been articulated as beneficial to both individuals and society when reduced to common practice in the marketplace. When embedded in complex social behaviors, such as business and economic activities, it has become a doctrine such as articulated by Adam Smith in his *Wealth of Nations*, published in 1776. In this context, self-interest is asserted as the most effective means to productivity that enjoys pride of place in the value network of a society primarily defined by commercial relationships. Productivity and the advantages that result are best accomplished through the exercise of self-interest rather than altruism as a competing principle of action. According to the Smithian view, both enlightened and unenlightened altruism are inadequate for achieving productivity or prosperity and have been shown historically to be used as disguised justifications for incremental manipulation, exploitation and control by those in power.[7]

Although Smith advocates the generous exercise of benevolence in guiding one's moral life and actions in his slim book, *The Theory of Moral Sentiments,* the economic driver of economic self-interest captivated the attention of many readers operating in the marketplace. The appeal of Smithian self-interest became apparent in economic theory, business transactions, corporate cultures and more broadly in related areas of sociology, psychology, and historiography, along with the natural sciences. Its central message resonated with those seeking a consistent and plausible explanation for foundational motives in both human and non-human behaviors. In order to survive, all forms of life appeared to seek their self-interest and without it were less "fit." When self-interest is combined as essential for the best way of earning a living within the Darwinian survival of the fittest doctrine, the broad social implications of Charles Spencer became more compelling. In this Spencerian iteration, Darwinian forms of self-interest and broad explanations and justifications of self-seeking, self-serving and self-absorbed actions are offered to generate wealth, benefit society, foster survival of the fittest and contribute to personal well-being. The results of prosperity generated by these behaviors have also been given religious meaning and standing as evidence of worthiness for God's chosen ones.

In contrast to both the enlightened and unenlightened Smithians, including neo-Smithians, principles such as equality and civil rights of individual humans living in a social context give rise to a broadly based humanitarian ethic. In undiluted forms of prescriptions for human behavior, humanitarian models have supported actions ranging from specific types of altruism to systematic subservience of self-interest and preferential treatment of others. When translated into formalized value networks, modified humanitarian ethical models have produced legislation, including limiting working hours for children and employees, health insurance, social security and distribution of food and healthcare to the poor. In its extreme versions, equality has become egalitarian such that resources are taken from those who possess them by the coercive power of the state and by military force. Programs of social welfare to spread the wealth become foundational for relief efforts in the name of distributive or social justice and under the banner of charity or the protection of collective civil rights.[8]

Despite the apparent benevolence embedded within a humanitarian ethic, its tangible expressions have been subject to generalized and specific critiques. A critique of rich countries' food distribution to poor countries, for example, suggests that these programs are often disguised expressions of self-interest. Specifically, an analysis published in 1986 states that these food relief programs are actually

> a means for developing markets, for helping agribusiness, for gaining a stranglehold on the policy decision of needy governments and for promoting U.S. foreign policy and military goals.[9]

More specifically, major legislation for food distribution in the form of Public Law 480, passed in 1954, makes the motivation of this act explicit. Susan George points out:

> Public Law 480 (Now also called the Food for Peace Law) [was] passed in 1954 with these stated purposes: An Act to increase the consumption of the United

> States agricultural commodities in foreign countries, to improve the foreign relations of the United States and for other purposes. The Congress further specifically declared that one goal was "to develop and expand export markets" for American products.[10]

A basis for these claims is found in the results of this Act that may appear on the surface as altruistic work. The relevant facts indicate that in 1954, Japan, for example, received nearly $400 million dollars in food pursuant to the Act, but over the next 20 years, Japan had purchased over $20 billion worth of food from the U.S. By 2005, Japan purchases over $2 billion in food products from the U.S. annually. In addition, the food given and purchased substantially changed the diet of the Japanese people such that specific cancers, heart and cardiovascular diseases significantly increased in Japan as U.S. food was consumed. An additional implication of the food exporting programs to any country, including Japan, is that the U.S. food is higher in overall fat, sugar and salt content that in turn, creates cravings for these foods, thereby expanding the markets for unhealthy foods.

These critical issues involved with real verses ostensible values of an ethical system appear in multiple permutations and take on added dimensions when embedded in political and social agendas. Within value networks found in entities that espouse a communitarian social values systems, key terms such as "good," "right" or "true" are defined within the context of the whole society.

Common good, the human community and the global village are also terms that are often found in communitarian ethical systems. Within the context of these systems, the interests of groups are viewed as having precedence over individual interests that are achieved by self-interest. On this view, various groups are thought to represent a cohesive and coherent value system that permeates the group and defines it in an identifiable manner. As such, a specific group holds itself out as one that will conduct its actions in predictable and reliable ways consistent with the value network of that group. Individuals within the group are expected to adhere to the internal regulatory forces and duties imposed to modify individual behaviors in deference to the group value system.

Models constructed on communitarian value systems including implied guidelines, rules and policies appear to create an alternative and a curb for unbridled individual self-interest. However, even at the threshold level, the communitarian ethical models are open to substantive questions including begging the questions of criteria for membership in a group, the right of a group to preempt individual rights and the balance of powers in regulating the actions of groups. For example, racial affiliation as an accident of birth has become a relevant rather than invidious criterion by which inclusion in a group is decided. As a member of a specific group, an individual may claim and assert rights based on group values that may augment and/or create additional prerogatives for the individual group member. When translated into political policy, membership in a group then grants individuals within that group access to various types of preferment such as affirmative action in education, minority employment or welfare emoluments.

When made integral to the structure and functions of corporations, communitarian value systems have given incremental powers to corporate groups, even though the legal status of a corporation has traditionally been defined as a person. In addition to affording additional powers to an individual directly and vicariously, values within a corporation give rise to a tangible value network in which the individual member becomes embedded. For the individual, continued habitation within the corporation requires acquiescence to the demands or obligations of that value network. The individual opposes that network at ones own peril and is similar to the citizen in Thomas Hobbes' Leviathan, who gives up individual prerogatives for "gain, security and reputation" provided by a strong group controlled by the coercive power of the sovereign.

These issues and competing views that have laid claim to incorporating authentic values and valid ethical systems are integral to decisions concerning technology development and the environment. Innovative technology development and commercialization inherently produce concerns regarding their affects on the environment due to the direct operation of the technologies and to the indirect contamination caused by their operations. The development of the chainsaw technology, for example, allowed for deforestation and destruction of ancient timber stands unprecedented in human history. Massive clear-cutting for timber harvests and mining, using ever-larger versions of the chainsaw technology, has allowed depletion of forests worldwide during the 20th century alone by nearly 35%. Ancient rain forests along with hard-wood trees throughout the world have been destroyed by nearly 50% during the same century.

Indirect affects are incalculable but include destruction of human and animal habitats with resulting death and displacement, contamination of water, soil and air by the toxic residue of the two-stroke chainsaw engines, along with global warming and acid rain. The two-stroke engine also used in motorcycles, ATVs, jetskis, lawn implements, boat motors and snowmobiles spews out 1 gallon of toxic fumes, carbon monoxide and petroleum residue for every 5 gallons of fuel consumed. On balance, the chainsaw cutting technology has wreaked more environmental degradation in one century than during the prior 8,000 years since saws were used in the Western World. While the chainsaw technology may seem to be a useful, efficient and beneficial device, the multiple direct and indirect consequences of this technology inflict far-reaching harm on people, animals, places and things.

Long lists of technology applications, increasing in numbers and widespread results especially during the 20th century, cause severe environmental degradation. While economic motives drive many of the decisions to develop, produce and distribute noxious technological devices, primitive needs for food, clothing, water and shelter generate chronic demands to use these devices. Short-term needs for shelter and cooking fuel in many underdeveloped countries without alternatives become justifications for deforestation with chainsaw crews. As suggested in a study of public health and environmentalism,

> faced with a choice of being unable to find fuel to cook rice and cutting the last remaining forest, it is usually argued that anyone near starvation will cut the forest. Here the conversation is thought to end.[11]

Where this conversation is thought to end, the origins of human thought and behaviors relevant to crucial choices actually begin. Decision-making regarding the development of technologies and their affects on the environment begins within the context of human survival and the choices that appear to enhance it. The choices made during human evolution, including those made by tribes, societies and cultures that developed, marked the beginnings of multiple value networks. These established hierarchies of values and choices defined the basis for orthodox thought and behaviors within social groups that became the dominant *mores* reflecting the expected value network memes and norms within those groups.

Chapter Two

Values in Contemporary Ethical and policy Debates

The relationship between the natural world, including the language domains that deal with natural facts and the realm of values has traditionally been opaque. Its opacity has been somewhat modified during the 20th century, especially since the articulation by G. E. Moore of the "naturalistic fallacy." According to Moore, all statements regarding values including those commonly used in ethical discourse are "non-natural entities" in that they are not generated from sensory observations. As such, values on this view are not derived from observations of natural phenomena nor can they be defined from observations of nature. Moreover, Moore argues that primary values cannot be defined at all but are to be considered as simple, discreet and unanalyzable entities. "Good," for example, as a common value term cannot be derived from a descriptive proposition regarding nature but is to be understood only in terms of its own inherent meaning. Thus, any ethical prescriptive statements recommending certain choices or behaviors based on their attribution of good cannot be based on descriptions of natural fact. In a word, an "ought" (prescriptive) proposition cannot be derived from an "is" (descriptive) proposition.

Despite these caveats, Moore recognized that it is possible to determine what are and have been certain types of "good" obtainable by humans. Although the fundamental element of good as the value basis is irreducible, it, nonetheless, can be recognized by "intuition" as having an affinity for bonding with a broad range of human behaviors. After binding to these behaviors, the element of good can become part of synthetic statements wherein a behavior is described or prescribed in a proposition. As descriptive, such a synthetic statement may become a subject of study for a commensurately broad range of disciplines. When part of a prescriptive statement, a value such as "good" is conveyed as an "ought" wherein it is recommended as an ethical imperative.

Even though a basic value such as "good" may not be derived from empirical data and is irreducible and indefinable, its adherence to complex wholes can be an object of empirical study. While the inherent "good" may only be accessed by intuition, its existence and quantity may be ascertained through various types of analysis including the scientific method. Furthermore, it is possible to conclude that certain complex wholes have a certain amount of that value inherently or in terms of their consequences. Within this framework, the elusive essence of a value such as good is not sought or captured by any means other than intuition; but those complex wholes bonded to goods "obtainable by humans" can be understood in their multiple forms by many types of knowledge.

Since these complex wholes or organic unities are discernable entities amenable to analysis, it is possible to study values binding to these unities of various kinds. For example, it would be possible to study, analyze and examine the proposition: "It is good to feed the poor" by subjecting it to a wide range of analytical methods. Moore, himself suggests that a useful method for analysis would include a hypothetical case wherein it would be better for the world if it existed alone as a maxim for human behavior. Furthermore, good as an indefinable term in itself can be further elucidated as part of an organic unity and, in turn, can be arranged within a hierarchy of values. Consequently, another organic unity conveyed in a proposition such as: "It is good to preserve the forests" may be ranked the same as, below or higher in value than the former maxim.

Although Moore's view of ethical behaviors guided by the primary value of good has been subjected to sharp criticism, it is instructive to determine the actual basis of values themselves. The fundamental issues of epistemology are also illustrated in looking at the ways values are claimed to be knowable. Intuition of the good, the answer given by Moore is, of course, open to epistemological critiques from many sides. But the seemingly simple conclusions still allow for fruitful inquiries into ways in which peoples from many eras and locations have perceived values including good. These values and perceptions have been bonded to behaviors, thereby forming organic unities that can convey ways of living. A careful record of these forms of life can provide understanding into the values that are embedded in the activities within groups of humans.

By studying the records recounting ways of life, the hierarchy of values within groups can be identified and analyzed by various ways and means. Moore himself opined that humans have achieved the most pre-eminent goods: "...personal affections and aesthetic enjoyments include by far the greatest goods with which we are acquainted." Moore's recognition that aesthetic enjoyments possess the value of good at the highest qualitative and quantitative levels suggests that the value good also forms the basis for aesthetics. Although it is not clear within Moore's discussion of values how aesthetics and ethics are related, it is clear that the value "good" is foundational to both.

Apart from the epistemological questions that surround the assertion that good can only be known by intuition, ranking aesthetic enjoyment as pre-eminent within a hierarchy of human attainment is useful to assess the origins of values and their roles played in decision-making. This includes judgments, choices and behaviors that have affected the environment for millennia. In fact, the importance of aesthetics relating to nature has been increasingly articulated at the end of the 20th century by those interested in environmental topics. Allen Carlson, for example, speaks to these issues in the following:

> The natural environment in so far as it is untouched by humans, has mainly positive aesthetic qualities; it is graceful, delicate, intense, unified, orderly, not dull, bland, insipid, incoherent, chaotic. All virgin nature, in short, is essentially aesthetically good. The appropriate or correct aesthetic appreciation of the natural world is basically positive and negative aesthetic judgments have little or no place.[12]

This view, articulated by Carlson at the end of the 20th century, represents an alternative and a fundamental basis of value. While this aesthetic basis of value is distinct from others that are foundational to ethics, there are close similarities that connect both. One similarity is the common denominator epistemological way of discerning the basis of value. For Moore, the most fundamental value of ethics, good, was unanalyzable and knowable only by intuition; writers such as Carlson also find their origins of value by invoking an intuition of aesthetic value located in nature untouched by humans.

Nature without human intervention is to be understood as having the highest positive aesthetic value without additional analysis or proof. Both Moore and Carlson as spokespersons for their widely held views identify intuition as the primary means of grasping the origins of value. Both take an intuitive awareness of the values underlying their traditions of thought as essential, inherent and fundamental.

For Moore, further analysis and evaluations of ethical choices or behaviors are still possible when the good, as a primary value, binds to form organic unities. Also, a hierarchy of choices or behaviors can be constructed using a broad range of these organic unities as building materials. It is on this basis that evaluations of values can be made insofar as certain choices and behaviors deviate from the highest and best values, such as "human affections" and "aesthetic enjoyments." Similarly, the aesthetic position of unmodified nature developed by Carlson is the framework and standard against which environmental choices and behaviors are judged.

At the outset, Carlson's environmental aesthetics appear to locate the origins of values in a place where they can be unmistakably recognized by one's intuition. Environmental values are those that are manifested initially in nature untrammeled and untouched by humans. Rather than using good as the fundamental value discerned by intuition, virgin nature becomes the basis of the highest positive aesthetic qualities. This position is clearly stated by Carlson in his assertion that "each natural thing…has substantial aesthetic value and little, if any, negative aesthetic value."[13] On this view, the gold standard by which environmental issues are judged is nature untouched, hence not contaminated by human modification of any kind. As a necessary logical conclusion, any modification or change of nature brings about consequences that lessen the quality and quantity of aesthetic values.

On this level of argument, an ethical values standard and an aesthetic values standard are seen to reinforce each other. Ethics based on consequences, for example, such as classic utilitarianism, would hold that actions that lessen value as a consequence are inherently wrong. This would be the case for both rule and act utilitarianism and for those who calculate the greatest good for the greatest number in the long run. Whether or not the coalescence of ethics and aesthetics on this basis is persuasive, the reasons given for inherent positive aesthetic value of virgin nature are worthy of consideration.

The primary reasons given for the substantial aesthetic value of untouched nature are that nature is taken as orderly, harmonious, unified and regular. These are the characteristics of nature submitted by Carlson that make nature aesthetically the gold standard and the ultimate basis of value. A position identifying the origins of value in nature without environmental alterations as a beginning standpoint is decidedly non-anthropocentric since the basis for values and evaluation is apart from human intervention altogether.

By placing the basis of environmental value in virgin nature, Carlson's school of thought replaces a broad range of values built on anthropomorphic foundations. These foundations have supported a variety of views that share the common denominator of mankind as the standard and arbiter of values. As such, the beginning reference points and the ultimate source of values in an anthropomorphic system are human-centered, not nature-centered. An extreme expression of this form of anthrocentrism would hold to the proposition that whatever is beneficial to humans is good and embodies the highest aesthetic values. This position disregards the negative outcomes and consequences for any other forms of non-human life or their environments. Radical anthrocentrism in contemporary environmental ethics and aesthetics emerges from an ancient tradition articulated in the dialogues of Plato. Protagoras, as its spokesperson summarizes radical anthrocentrism in his statement that "mankind is the measure of all things."[14]

These fundamental issues appear in various forms and permutations within both theoretical debates and those that focus on practical policy agendas. In the broad area of environmental issues affected by ethical and aesthetic value discussions, relevant conceptual constructs and policy formulation needs are seen to intersect. One such example is found in the Maine Woods program that is the subject of Joshua Haddock's case study entitled "The Great Misguided Wilderness Debate. A Goal-Oriented Approach to Environmental Policy Deliberations." The program involves 3.2 million acres of wilderness located in the State of Maine that has as its primary goal to preserve native fauna and flora within the landscape and eco-system of a "vast cove of wilderness." Within the context of this ongoing project are found the convergence and conflicts that develop whenever broad concepts, no matter how venerable, collide with the practical urgencies of accommodating conflicting interests.

The Wilderness Acts of 1964 set forth the broad conceptual statement of wilderness as a teleological-driven goal to be achieved. In part, it designates

> a wilderness, in contrast with those areas where man and his own works dominate the landscape, is hereby recognized as an area where the earth and its community of life are untrammeled by man, where man himself is a visitor and does not remain.[15]

Despite its *prima facie* appeal to many, the fundamental wilderness concept articulated has been subjected to a broad range of robust criticism based on several independent and overlapping premises. These include those stated in conclusionary summaries that characterize the basic wilderness concept as "...ethnocentric, androcentric, phallocentric, unscientific, unphilosophic, impolitic, outmoded [and] even genocidal." [16]

While these conclusions may be supported by logical and rational arguments, the somewhat inflammatory rhetoric serves to illustrate the profound differences between the parties of interest. It also shows the language distance and lack of common grounds for productive communication between those relying on concepts rather than the percepts of practical exigencies. Without a shared universe of discourse for meaningful discussions, the environmental conceptualists and the policymakers are as separated as C. P. Snow's scientists at Cambridge were from the liberal arts schools. As cited from *Environmental Pragmatism* by Light and Katz in Haddock's paper:

> The intramural debates of environmental philosophers, although interesting, provocative and complex, seem to have no real impact on the deliberations of environmental scientists, activities and policymakers.[17]

Despite firm and even entrenched beliefs regarding the ultimate basis of inherent, self-evident or significant values supporting certain claims, a widespread disagreement continues among the disputants. This unsettled condition persists and is even exacerbated as the stated values of each disputant are more dogmatically stated. The result, as Haddock indicates, is that there is an increasing divergence among the principle disputants, thereby precluding productive problem solving. In particular, the rarified air of those producing conceptual rhetoric has failed to lift the practical policymakers to their level of discourse. As a result, those segments of the disputants are ignored as irrelevant to the practical and pragmatic work of policymaking.

This long-lasting stalement among the disputants in the Maine Woods example highlights a profound quandary generated by those seeking intuitively understood values. While seductive as an epistemological basis for values foundational for ethical, aesthetic and legal decision-making, a fundamental weakness of this position is exposed when applied to practical policy formulation. What is intuited as inherently good or right is frequently countered by alternative intuitions, some of which are in direct opposition to one another. Even when bound to other elements to form organic unities as proposed by G. E. Moore to form more complex propositions, these organic unities do not appear to be useful arbiters for practical decision-making. Those espousing them are still considered suspect in their concept formulations, even though more complex propositions are constructed.

The quest for indisputable foundations of value, as self-evident as Descartes' *indubitandum* (that which cannot be doubted) and the *cogito ergo sum* (I think, therefore I am) has resulted in confusion and conflict rather than in accord and consensus. In the case of the Maine Woods illustration, the points of conflict are multiple wherein the irreducible values of untouched nature are juxtaposed with the felt need to formulate practical policy such as multiple uses that are contrary to the complex organic unities of the wilderness concept.

Due to these points of conflict rather than accord, parties to these kinds of disputes find themselves embroiled in seemingly irreconcilable differences. Despite robust debate, analysis and strenuous efforts to persuade the opposition, appeals to reason, logic, empirical data, intuition and common sense for finding common grounds of bedrock values have largely failed. Failure in these efforts has translated to frustration, suspicion, and hostility and has created critical stalemates in managing dynamic eco-systems. Regardless of the conflicting points of view, the environment itself as the final arbiter of human wisdom or folly continues to reflect the outcomes of these quarrels waged with languages generated from various Towers of Babel.

Chapter Three

Environmental Outcomes of Goal-Directed Concepts and Actions

Despite carefully constructed arguments that attempt to build fundamental bases for values in ethics and aesthetics, unresolved issues continue to proliferate when applied to human actions. These include quandaries arising when value concepts collide with ongoing needs to form policies that guide human behavior and have direct affects on other forms of life and their environments.

In addition to providing a useful case study in the Maine Woods debate, Haddock also suggests a way to abate some of the conflicts experienced in these kinds of conflicts. The primary way proposed is one articulated by Douglas Walton in his book *The New Dialectic: Conversational Contexts of Arguments.*[18] Deliberation, as Walton states it, is a method and means to clarify common goals such that the participants are able to continue a discussion with an end in view. The outcomes of this deliberation are to arrive at agreements regarding appropriate action policies that would lead to those goals. Achievement of goals would, according to this approach overcome many of the obstacles experienced when the disputants engage in their continuing debates. Moreover, common goals identified at the threshold of the discussions would establish objectives to be obtained regardless of the differences in values, principles or language used in the process. As there may be many means and pathways to summit a mountaintop, the routes chosen become subordinate to the common goals.

In fact, the trek to the mountain top analogy is apt since Walton suggests at the outset that there are two major premises involved in the simplest formulation of his argument. They are the "goal" premise and the "means" premise wherein the latter states ways in which the desired goals can actually be achieved. Both premises are set forth within the context of practical reasoning that is considered to be "inherently teleological."[19] As teleological, the process with all of its elements is essentially driven toward bringing about a result that is defined in terms of stated goals.

Once the goals and means premises are agreed upon, Walton's methodology involves the generation of additional premises relevant to the desired outcomes. These include the "selection premise," the "practicality premise" and the "side effects" premise, all of which are situated within the context of language that is both practical and reasonable. With these elements in place, a productive dynamic is thought to develop wherein the dissonance, conflict and disagreement between specified policymaking goals and the conceptual ideologies can be greatly ameliorated if not entirely abated.

However, the broader dimensions of certain fundamental issues involved such as teleology itself appear to require further discussion prior to evaluating the resolutions offered by Walton's approach. Teleology as an explanatory device has been employed in most every academic discipline from antiquity. Ancient writers in the Western world, including Plato, Aristotle and their followers, provided teleological accounts of phenomena in most every realm of inquiry including biology, cosmology, ethics, aesthetics, psychology and theology. The underlying and foundational role of teleological explanations for both inanimate structures in the universe and life forms has been a perennial quest of thinkers over many millennia. A discerned goal amid seemingly chaotic behaviors of various life forms provides an explanatory framework, a basis for understanding and the power to predict future activities.

For inanimate structures making up the metes and bounds of the environment for the various forms of life, teleological explanations offer plausible relationships that may exist in the great chain of being. This great chain may be forged from unique and unlike links, yet may be seen as binding the universe itself together in macro-and micro-environments. Teleological explanations within a macro-world environment have taken the form of cosmological views that impart purpose and meaning to an otherwise confusing collection of objects controlled by capricious and arbitrary forces. Animal and human behaviors have been explained in teleological terms, despite the complexities of human actions and the perils of anthropomorphizing by foisting human explanations on animals. On a rational model such as Platonism, human reason is thought to provide the checks and balances needed for controlling passions, drives and desires that Plato characterizes as unruly and willful horses. Plato viewed reason's *telos* or purpose to primarily govern and give direction to otherwise irrational and unmanageable passions or personal demons.

Aristotle, one of Plato's most illustrious students, sharpened the teleological focus even more by identifying that the concupiscent and irascible appetites could only be managed by the cultivation of virtue under the direction of reason. Human virtue, defined by Aristotle as a state or habit of character lying is a mean between excess and defect relative to ourselves and as wise persons stipulate, needs to be developed by repetition. The final outcome of this repetitive discipline allows a human being to achieve a personal *telos* or purpose. In this context, Aristotle makes an astute observation regarding the compatibility of moral virtues with our human nature. His analysis results in the conclusion that our human nature does not of itself produce virtues but does not "oppose" them despite that habitual repetition is required for the virtuous life to occur. For Aristotle, the individual and collective *telos* for humans was *eudaemonia*, the ultimate goal of ethical human actions.

Although eudaemonia has been commonly represented over the preceding 2,350 years as happiness, the Greek phrase *eudaemonia* actually means controlling one's demons. This meaning, based on both the denotation and connotation of Aristotle's writings in the *Nichomachean Ethics*, produces a distinctly different *telos* ascribed to human ethical actions than does happiness. Happiness, a derivative from the French *happe* means a lucky chance occurrence and is distinctly different than the sturdy and purposeful conquest of one's demons by the rigorous practice of virtue. Within a collective context, both Plato and Aristotle utilize teleological explanations to explain group behavior, including the collection of large organizations such as the city-state. When city-states are formed, the individual, family and tribe values consolidated as purpose or goals become subservient to the dominant teleology of the city-state. The primary reasons for formation of these larger city-states were ostensibly to provide increased security and prosperity for those within the larger groups. Within an ideal city-state, individual, family and tribe *telos* would combine to strengthen the collective body as do the purposeful functions of body parts sustain organic bodies.

However, both Plato and Aristotle realized that the realities involved in these processes were often in sharp contrast to the *aretean* ideals of harmonious syntheses and the construction of orderly and equitable societies. Even though the over-riding *telos* of the city-state may be defined and understood, compliance to its normative dictates in the form of ethics and law required a multiplicity of ancillary structures for enforcement. Although Plato recognized the disparity of individual abilities or capacities by stating in the *Republic* that there are bronze, silver and gold classes of humans, his confidence in reason's capabilities shared by all the classes to control individual and group passions made possible a functioning city-state *telos*. Despite his credence in human reason to discern and sustain a group *telos*, Plato still looked to a city-state ruler who was to be a benign philosopher king. This Platonic tradition has manifested itself over several millennia and often has taken the form of utopianism in its many permutations. In such a utopian city-state, while there may be limitations on size such as Plato's limit of 4,000 individuals or care of children by professional nannies, the collective *telos* is an equally conceptual ideal operating within the body politic.[20]

Both the Platonic and Aristotelian traditions have continued to express themselves directly and indirectly over several millennia in serious discussions regarding conceptual ideals and applied policies on a wide range of topics. Many concepts are derivative from the Platonic ideals formulated in the seemingly rarified air of the good, the true and the beautiful. From the same Platonic traditions come practical ways of structuring human lives and their behaviors in the carefully constructed social and political models found in the *Republic*. Similarly, the Aristotelian tradition is equally broad in its scope and has been the foundation upon which other venerable traditions have been based, including the extensive work of Thomas Aquinas. Whether standing on its own or assimilated into a Christian Thomistic world-view, Aristotle's legacy also extends to the lofty atmosphere of concepts. However, for Aristotle the conceptual world is to be reached through concepts based on percepts. The Platonists remember their participation in pure ideas temporarily forgotten by the trauma of birth while the Aristotelian needs to laboriously learn from the sense data of experience.

Despite the fundamental differences in terms of epistemology and metaphysics with additional accretions in the Thomistic tradition, teleological explanations pervade these schools of thought. It is especially noteworthy that the elements of these traditions have also been incorporated into the work of modern thinkers and into contemporary debates. John Locke, whose writings influenced the American Revolution, brought Platonic ideals into contact with human minds that were originally a *tabula rasa*, or blank slate. The combination of immutable ideals translated into rights such as life, liberty and the pursuit of property or happiness constructed with the percepts of experience resulted in a consolidated national *telos*. It was on these bases that a focused collective purpose could be structured on a conceptual level and on a practical policy level to mobilize both collective ideals and regiments of troops marching into cannon fire.

As traditional teleological explanations are

translated into lists or litanies of rights, the basis and scope of the discourses change commensurately. Similarly, when rights are derived from teleological explanations, the focus, the means and the ends also shift to encompass an increasingly wide array of issues. In a monograph presented at the 2005 Applied and Professional Ethics Association, the author addresses the complex topics of parenting and reproductive choices that illustrate the expanding uses of teleological explanations.[21] The discussion of parents as stewards shows the implications of a rights-based argument as applied to the parent-child-family-community relationship. The author argues that a rights-based paradigm such as proposed by John Locke does not give adequate account of the needed element of stewardship within the context of procreative choices.

Central to the discussion are the uses of pre-implantation genetic diagnosis (PGD) and assisted reproductive technologies (ART) in the processes of producing a child. By utilizing the means of PGD and/or ART, the parent becomes increasingly implicated in affecting the actual *telos* of the child with all of those ramifications for the family and the community at large. In this way, the naturally occurring *telos* of offspring transmogrifies into one designed by parents and directed toward desired outcomes. The concern emerging from individual parental selections is compounded when a standardized set of selection criteria may result in the contrived creation of a large "class" of children engineered with prescribed traits.

The gold, silver and bronze classes suggested by Plato could be further broken down into myriad subsets as a result of technological control over procreative *telos* by choices advanced as inherent or inalienable rights. When these putative rights are extended to include the broad concept of individual freedom or liberty, parental choices then incorporate ancillary rights such as entitlement or self-actualization to further reinforce an agenda of teleological design. Medical science and technology are enlisted to select desired characteristics of progeny and to cull out the potential defects of gestation either before conception by PGD or after implantation by inducing fetal demise using the means of abortion.

While animal breeding by natural selection or traditional trait breeding of livestock have been accepted and performed for millennia, these techniques have been limited when applied to human breeding. The over-riding questions regarding quantity and quality of human reproduction have been a topic of debate since antiquity, but the answers given have varied widely over the centuries. Limitations on quantities of offspring have affected quality both for animal and human breeding. This statistical reality has been brought into sharp focus wherever governmental policies penalize reproduction of more than one child per breeding pair in countries such as China. Under these restrictions, there is additional concern to insure the quality of a single child by the use of PGD and other means that allow breeding for specific traits.

Conversely, policies that restrict the use of contraceptives and abortion to control reproductive quantities have a direct impact on the statistical quality of offspring.[22] Ethical tenents, embedded in religious institutions and made into moral imperatives or translated into legislation and/or court decisions, have a profound impact on reproductive decisions. Historically, those institutions of authority that have focused on empire-building and global colonization have emphasized quantitative multiplication, whereas the institutions directed toward stability and maintenance tend to encourage quantitative restrictions in the interests of population control and qualitative improvements of phenotypes within populations.[23] Since there is discernable interaction among individuals including their social groups and their environment, there are commensurate reciprocal impacts on the *telos* of each. Because of this continuing relationship, the *telos* of the individual person, whether naturally occurring or modified by technology, also creates needs and wants that place measurable demands on the resources of a specific environment. Generally, increasing the quantity of individuals within a population creates a sharply increased demand for the necessities of food, clothing, shelter and healthcare as minimum requirements to sustain life.

Critiques of methods, including the technologies involved with PGD and ART that affect both the quantities and qualities of progeny, often argue that these methodologies are an inadequate basis for reproductive decision-making. Moreover, parental decisions made on these bases often favor the parents' own interests rather than the child's and may arbitrarily or artificially control the destiny of the child and its *telos* using a parental rights-based justification. This way of proceeding with reproductive choices according to its critics does not recognize the inherent rights of the child pursuant to natural law and it fails to recognize ancillary considerations necessary in the parent-child relationship.

These reproductive decisions have a direct and significant impact on the major elements of the quantities and the quality of a population; and, the population itself is a primary factor in assessing statistical impacts on the environment and the eco-systems of the earth. Teleological explanations and ways of establishing individual and collective goals have been useful means in a broad range of disciplines. However, the ambiguities, resulting from teleological explanations used in environmental issues to arrive at common goals, make it necessary to evaluate additional methods to resolve the impasse between the environmental conceptualists and the practical policy demands of applied environmental ethics.

Chapter Four

Reproductive Choices, Procreative Norms and Their Environmental Impact

The impasse between the environmental conceptualists, *vis a vis* the policymakers, bears a distinct similarity between those advocating the use of biotechnology devices in reproductive choices and those who do not. Those opposing the use of biotechnology means, including PGD and ART offer both negative and affirmative arguments based on a wide range of natural law premises. Negative critiques of biotechnology assisted conception have encompassed specific deficiencies of this choice such as those suggested by Brent Waters. For Waters, devices such as PGD and/or ART fail to recognize other essential issues relevant to the relationship among parents, children and community.[24]

In addition to a wide range of specific conceptual criticisms, a broad agenda is found that urges the application of alternatives to guide parents' fertility behaviors. Such an agenda uses an alternative vocabulary derived from the traditional lexicon of natural law theory and religious practices. Terms such as covenant, co-creation, familial love, beneficence, mutual sacrifice, justice, compassion, stewardship and love are central to the teaching of natural law theory and practice. Words such as these convey a conceptual dimension that is reflected in a cohesive language that supports a way of life and decision-making in sharp contrast to that proposed by biotechnology advocates.

In the extreme versions of these concepts and at the outer limits of their applications, the specters of irreconcilable differences are seen. The conflict between the environmental conceptualists and the policymakers in the extreme is a conflict between partisans that demand nature to be untouched and those wanting to alter nature according to their designs. "Untrammeled nature" is the highest value for those wanting to preserve natural wilderness while the "highest and best" use of the environment is planned alteration of natural wilderness according to governmental and commercial policies. The quest to arrive at common grounds based on the identification of common goals continues unattained. Similarly, the debate regarding reproductive choices using biotechnological devices and procreative alternatives has become increasingly strident especially when parties to the debate endorse the most extreme expressions of their positions.

Due to the profound and irreversible impact of a population on any environment, relevant issues affecting both the quality and quantity of reproduction need to be evaluated. Human populations, not unlike other animal populations, forage for food, depend upon water sources, seek and fabricate shelter, and compete for territory. In addition to these rudimentary appertences of life, humans desire an extensive list of additional products and services, all of which require additional resources to produce, distribute and consume. The carrying capacity of the commons applies to animals and humans alike, although the exponentially increased demands to satiate human consumption wants cause depletion of the commons at a greatly accelerated rate.

Under these chronic conditions, the competing claims of those recommending the use of biotechnology in reproductive choices and those proposing alternatives are to be assessed. Arguing either of these competing positions as only a conceptual debate without actual application to a world of limited resources would restrict the debate largely to arbitrary definitions, semantics and logical syllogisms. Such a contest would focus on tautologies, internal coherence and logical consistency rather than the rigors of showing actual correspondence of concepts and percepts to a world of relevant facts.

Those positions that offer alternatives to reproductive choices using various biotechnology devices to assist in the outcome, refer to concepts that are given coherence by definition. While some of those terms have both traditional denotations and contemporary connotations, other concepts are constructed from "terms of art." As terms of art, their definitions are derivative from both explicit and implied meanings within various contexts.

The concepts that are derived from these terms, in turn, are often ambiguous or equivocal and frequently are without useful content when employed in applied policymaking discussions. When these concepts are fabricated from a network of contrived terms, a dialogue on both conceptual and an applied policy levels becomes increasingly difficult.

These difficulties are confronted when reviewing the theoretical concepts and proposed applications for considering alternatives to the uses of biotechnological assisted reproductive choices and behaviors. Such a representative alternative position advanced by Brent Waters, for example, uses terms of art, traditional terms and those that may appear contrived to opposing parties of the debate. At the outset of his criticism of reproductive technology, Waters assumes that procreation should take place within the context of marriage wherein that "bond of marriage" implies the existence of a "covenant relationship" among the parents, their child and the larger community. From this fundamental bond, Waters states, certain implied virtues emerge in which stewardship is identified as central to procreative behavior, including the ongoing interactions between parent and child. The virtue of stewardship with consistent procreative behaviors is combined into terms of art referred to as "procreative stewardship" which Waters defines as follows: "Procreative stewardship is based on a general precept that the means employed in ordering procreation should accord with and enable a more expansive familial love."[25]

This statement appears to make no general condemnation of biotechnological means such as ART and PGD or offer a *prima facie* prohibition of its use in procreation planning and preparation. However, their application to specific actions and behaviors further qualifies the legitimate uses of biotechnology means in procreation by allowing their uses only as a complement or facilitator for natural procreative behaviors. While these biotechnology devices are generally thought to be neutral as concepts, their actual uses within the context of procreation are limited by the conditions of procreative stewardship. Essentially, those means that run counter to the explicit and implicit definition of procreative stewardship are in violation of the normative operations of those virtues arising from adherence to its premises.

Although these alternatives as restrictions to reproductive biotechnology are advanced as sound concepts and useful in application, outcomes generated by the practice of procreative stewardship meet significant challenges when applied to "hard cases." These "hard cases" are encountered in a wide variety of reproductive areas and appear in a multiplicity of relevant health-related categories. One of the broad categories is termed teratogenic in that it is derived from its Greek etymological origin meaning "terrible birth." The cases that fall within the teratogenic category include all those births manifesting a spectrum of malformations ranging from a congenital anomaly through increasingly severe debilitating defects to those that are life-threatening throughout the newborn's lifespan.

Statistically, teratogenic events occur in approximately 3 percent (3%) of all human newborns and one-third of those have life-threatening conditions. As infancy develops into childhood, twice that number of medically significant defects are detected and nearly fifty percent (50%) of all children in hospitals are there because of prenatal malformations.[26] Although the specific causes of these malformations in humans considered teratogenic are not known in a majority of cases, it is estimated that approximately twenty percent (20%) are caused by genetic defects, five percent (5%) are caused by chromosomal aberrations and approximately ten percent (10%) are known to be caused by teratogenic agents. The result is that nearly sixty five percent (65%) of human teratogenic births are of unknown etiology. In terms of teratogenic agents that cause these malformations, there are fewer than fifty that have been shown to be definite causal agents in humans (Appendix A).

However, over nine hundred agents have been identified that cause teratogenic effects in many animal species. This disparity between animal-test findings and human testing is thought not to be human resistance to these agents but a sizable short-fall in animal testing results that are applied to humans. Regardless of this experimental gap, the quantity of malformations and the severity of the teratogenic results constitute significant numbers of "hard cases" that challenge those opposing biotechnological devices in reproductive choices. Since there is at least a thirty five percent (35%) informational basis, those technological means that utilize the information relevant to reproductive outcomes provide that degree of statistical reliability. In specific cases, of course, in all three etiological categories there is nearly a one hundred percent (100%) probability of making a correct choice that is based on incontestable evidence. At certain known levels of exposure, for example, Thalidomide or nuclear radiation can be predicted as teratogenic agents with near certainty. Similarly, in the categories of genetic defects and chromosomal aberrations, it is statistically sound to predict with a high degree of certainty that a teratogenic birth will result in a *conceptus* carrying a specific gene or having a confirmed chromosome aberration.

When parents have knowledge of any known teratogenic factors that are present during gestation, there is reason to suspect the possibility of a teratogenic birth. With increments of relevant information that would include additional teratogenic factors, the statistical probability of fetus malformation increases sharply. In cases where these factors combine with accurate pre-implantation genetic diagnosis (PGD), the probability of specific malformations approaches 100%. If this correlation of near certainty is confirmed by the additional diagnostics of amniocentesis and/or ultrasound imaging, biotechnology information relevant to the outcome of gestation is virtually without question.

Over this entire statistical spectrum of information, the guidance offered by procreative stewardship is challenged and brings into question the viability of such teachings. While procreative stewardship rhetoric may resonate on a conceptual level and appeal to those already conversant with this type of language, the vocabulary from such a lexicon is often foreign to the uninitiated. This lack of recognition and use vocabulary often occurs in many theoretical discussions where concepts are the pathways taken to common grounds. Brent Water's caveat regarding the use of reproductive technology, for example, takes the form of "engineering our grace through a baptism of technological rather than supernatural means." In addition, he cautions the use of reproductive technology in that it "commodifies" the future child, inhibits spousal cooperation and may prevent an “embodied conception” of procreation.[27]

Despite the creative crafting of traditionally theological terms, such as baptism with biotechnology references, the resulting admonitions remain unclear even as statements within conceptual discourse. The contrast between technological and supernatural creates an ambiguous if not misleading dichotomy between human intervention and the workings of a supernatural being. If Waters means that human calculation of a potential teratogenic birth should be abandoned in deference to divine intervention, human rationality itself is brought into question. Waters' answer to this broad question on the conceptual level is that procreative stewardship inclines parents to accept potential teratogenic births rather than access biotechnology information.

The extreme iterations of this position include the suppressed assumption that human intervention is in conflict with divine will that has primacy within the context of procreative stewardship. On this level, the underlying assumptions of the advocates of "untrammeled wilderness" are similar to those who urge deference to supernatural means. For both, any intrusion by human devices of any kind alters the operations of unimpeded nature. Stewardship for both extreme conceptual agendas would mean leaving nature alone and accepting the operations of nature *in utero* and in the wilderness, absent the intrusion of any human intervention whatsoever. [28]

When these assumptions, rhetoric and norms are applied to a broad spectrum of hard cases, the consequences of this position become increasingly troubling. At the outer limits of both agendas, any human involvement or intervention in the workings of nature are considered objectionable and in violation of fundamental premises. For the Maine Woods purist, for example, the "wilderness" is to be inviolable, inalienable and indefeasible entirely without encroachment by humans (Appendix B). At this same degree of fundamentalism, the proponents of unqualified resistance to biotechnology use in procreative behaviors prefer supernatural intervention to the use of human biotechnological means. Within the parameters of these positions, common grounds based on common goals or acceptable means to these ends are conceptually impossible. This is essentially because the acceptance of a fundamentalist position regarding wilderness, development, procreation or reproductive choices logically precludes alternate positions along with their alternative conclusions.

Despite its seemingly paradoxical status on a *prima facie* basis, some degree of compromise between extreme fundamentalist positions may be found in the arena of "hard cases." This proposed reconciliation might be possible even though the areas of fundamental disagreement are inherently problematic and fraught with conflict on most every level of the discussion. Those advocating reproductive technology choices are able to show that reliable technologies currently exist that allow for a broad range of predictions ranging from sex selection through specific disease-trait identification. Additional traits including psychological orientation, IQ and body size predictions are also within the realm of probability during the first decade of the 21st century. With these biotechnological capacities available, a technology basis is in place to support the use of these devices in making at least some reproductive decisions.

In fact, the strength of these biotechnological capabilities allows some advocates of reproductive stewardship to recognize their use as permissible, but within the parameters of stated norms. Waters, for example, deals only with disease trait selection technology wherein such devices can be used to "prevent a severe illness or disability substantially incapacitating the parent-child relationship;" not violate the "embodied character" of procreation; or "guided by a practice of medicine submitting itself to the authority of the body."[29] Although all of these formulations are set forth as normative for disease-trait selection technology, it is not clear if these same norms can be applied to non-disease-trait issues. Despite this uncertainty, Water's reproductive stewardship analysis does recognize the utility of biotechnology to prevent serious diseases or incapacitating disabilities and condones its use to statistically prevent a teratogenic result of gestation. This prospect of producing a severely deformed or diseased child appears to be so compelling that even a procreative stewardship conceptualist is able to accept biotechnological means to prevent it.

In addition, there are plausible justifications for transforming permission into a moral obligation in using biotechnological means within the conceptual framework of procreative stewardship. Moreover, these same arguments go beyond the moral obligation to use biotechnological devices for the prevention not only of teratogenic outcomes but are to be used to prevent less serious defects. Under the justification of procreative beneficence, moral imperatives are applied to a broad range of non-disease or non-deformity traits.[30]

Apart from the objections implied from the application of Julian Savulescu's position based on procreative beneficence, the general agreement on a critical issue of severe defect or deformity brings procreative stewardship together with reproduction biotechnology advocates. Both sides recognize that a teratogenic birth needs to be prevented by available biotechnological means regardless of the conceptual disagreements that may be involved. In these hard cases, where there are and/or will be significant risks to the health of the mother or the fetus based on relevant diagnostic data, both sides to the debate are brought together to arrive at the same conclusion. Correct medical information indicating a high probability of a severely deformed or defective child creates compelling unanimity among informed participants to the decision that such an outcome should be prevented.

When the hard cases become softened, the conceptual and policy-driven arguments become increasingly divisive. There is more general agreement when a high probability exists that a child will be born deaf and/or blind due to rubella contracted by the mother; when there is a high probability that the child will be asthmatic, the debate becomes especially strident. When the issues involve low probability for less serious diseases or defects or a high probability for non-disease traits such as body size, IQ, or appearance, the debate intensifies on both the conceptual and the policy level of discussion. On this level, social justice premises conflict with individual rights and the fiduciary duties of parents oppose public policies regarding health and education; the financially affluent's preferred genetic selections collide with the "have-nots" who must take whatever is bestowed upon them. Within the context of both hard and soft cases, issues directly and indirectly affecting the environment must be addressed and solutions found.

Chapter Five

Needs, Wants and Their Environmental Consequences

It is possible to achieve a high degree of agreement among competing conceptual and policy claims to settle narrowly defined hard cases regarding severe disease or defects predicted for offspring. Even the most radical proponents of extreme positions involved with the use of biotechnology in reproduction or procreative choices can find common grounds and solutions for hard cases. The statistically immanent prospect of a severely deformed baby demands decisive action based on primal imperatives, akin to a crying infant, and not based on conceptual calculations or prolonged policy conferences. These imperatives may lack the measured academic cadence of Kantian categorical imperatives but still compel the actions of practical reason. Despite the inability to fully articulate the underlying motives or rationale, consistent action is taken on a pre-verbal basis or from a post-linguistic springboard. Schools of thought from classical sources to the present are eclipsed by these primal imperatives. Specific obstacles, such as skepticism, religious scrupulosity, cynicism and rationalism dissolve in the face of urgent basic needs made manifest.

The urgency of these needs becomes the Cartesian

indubitandum that moves one to action and takes the perplexed to the place of Ludwig Wittgenstein's example of one human being observing the compound fracture of another person. The question of that person's pain is a moot point. In this context, "Doubting must come to an end," states Wittgenstein; further pondering or committee sessions would be a study in futility. Needs that command such specific actions have been traditionally based on accounts of human nature and its inherent set of requirements for individual and collective survival. Food, clothing, shelter and water have been the four essential elements necessary to sustain humans long enough to reproduce, if not to care for their progeny. Hundreds of thousands of years of human evolution passed before healthcare was collectively added to the list of basic needs in the 19th century; formal education was included as a basic need as late as the 20th century. [31]

Needs of these fundamental kinds give rise to their own sets of demands that require appropriate and timely responses in order to be met. Unmet needs create urgencies that drive decisive actions, whether directly or indirectly, and, if unrequieted often lead to various types of conflict. Even though stealing food, for example may be a criminal act, theft of food by a starving person is ethically permissible and can be shown to be a moral obligation on most every ethical theory. Even traditional and conservative ethical authorities have recognized the ethical imperatives inherent in meeting basic needs. Thomas Aquinas, an authoritative spokesman for a conservative and *aretaean* ethical system since the early 14th century, states in the final analysis that "...(real) necessity knows no law."

While strong cases can be made for the ethical standing of needs, it is not always obvious what constitutes legitimate needs in every context. Even when bright lines are drawn on agreed-upon lists of real needs, it is not an easy matter to determine the environmental impact of meeting or not meeting those needs. Moreover, the means chosen to meet an unquestioned need for food, for example, create their own set of problems for which solutions are not obvious. A collective decision to produce food without synthetic herbicides, pesticides or the use of motorized machines would cause consequences that may be globally catastrophic. Raising enough animals to produce the necessary fertilizers and to supply drayage power to cultivate food products for the world population in the 21st century would increase methane gas to toxic levels, decrease available grain supplies, promote over-grazing of pasture land and decrease crop yield.

When transportation is included in the list of basic needs, the food requirements to feed riding or carriage horses would cause starvation for the animal herds and their multiplying caretakers. A bio-mass fuel such as ethanol can produce 3 gallons from each bushel of grain allowing approximately 75 miles of travel powered by a 200 hp engine resulting in a multiple of 15,000 hp/miles. By contrast, the same 1-bushel of grain would produce horse travel of approximately the same distance but at 1 hp or a multiple of 75 hp/miles in contrast to the ethanol output of 15,000 hp/miles from the same quantity of grain. Even based on the short travel distances prior to the internal combustion engine era and adding in an additional 20% increase to the efficiencies of animal grain consumption, without additional energy needed to convert it to ethanol, animal power is extremely inefficient.[32] With a five-fold population increase in the United States since the introduction of the internal combustion engine, transportation needs could not be met without significant social and environmental disruption. Any incremental increase in transportation distances to meet the expanding needs of a swelling population would immediately create commensurate marginal utility and long-term deterioration in most every sector of society and the environment.

When lists of wants are lengthened worldwide even before basic needs are met, the profound negative impacts on the environment often occur dramatically and profoundly. In many cases, these crises are clustered wherein the dominoes of one segment of an eco-system begin a sequence of related events resulting in the swift and equally certain demise of eco-systems. Many of these precipitous chain reaction collapses are also irreversible. Acid rain falls on those seeking to meet their basic needs as well on those demanding their endless lists of wants to be fulfilled. Unlike the ethical standing of needs, wants are morally unprotected except by contrived rationales and bogus justifications. However, we have not produced a sufficient anti-acid to neutralize our collective environmental heartburn, a global condition caused by daily excesses.

The magnitudes of these interconnected crises are well documented and corroborated by a multitude of scientific studies. Relevant findings from these studies have been summarized in a four-year research study conducted by the United Nations Millennium Eco-system Assessment in which 1,300 scientists throughout the world participated. An essential conclusion of this study is that the entire globe has been degraded largely due to humans demanding wants rather than meeting basic needs. The inevitable outcome is the prodigal squandering of non-renewable resources and the pollution of every eco-system through this systematic rapine of all available resources. A conclusion to this ongoing predation and plunder, driven by the insatiable value network of consumerism, is being witnessed at the beginning of the 21st century and awaits its inevitable disastrous outcomes within this same century.[33]

Satisfaction of basic needs alone places substantial burdens on most all eco-systems even when a generous carrying capacity is present and when populations of humans and animals are stable. When natural resources of the commons are depleted by overuse or lack of replenishment, eco-systems rapidly deteriorate and lose their capacity to meet the basic needs for the sustainability of many life forms. Phenomena, including global warming, pollution and encroachment of humans into animal habitat, have caused disruptions in eco-systems that heretofore had supported bio-diversity. Wildlife biologists, studying fauna and flora in Yosemite National Park since 1904 have documented profound changes in that ecology with resulting displacement and depletion of animal populations. Despite increased wildlife and habitat protection in that eco-system, there has been a rapid decline in multiple species of animals. With the remaining species, there is notable stress on those populations such that the patterns of animal life have been disrupted and changed unalterably.

The survival of even small animals such as indigenous rodents, chipmunks, wolverine, ground squirrels and multiple species of birds that require only meager forage is increasingly doubtful. One hundred years after the legendary biologist Joseph Grinnel pioneered rigorous wildlife studies in Yosemite, the unmistakable signs and symptoms of species stress, depopulation and demise are being observed even in this controlled environment. Since small animal species, especially on islands or in larger eco-systems segregated from other eco-systems, are reliable predictors of species stress and collapse in macro-eco-systems, the recent biological studies in Yosemite are especially prophetic (Appendix C).

These studies are well-documented and conclude that the primary causes for the erosion of the Yosemite environment are those associated with global warming.

There has been a 9-degree Fahrenheit rise in ambient temperature in that region since Grinnel began his research in 1904. This dramatic temperature elevation has profound and irreversible effects on the flora as vegetation is stunted and withers earlier in the growing season. Even a few degrees elevation in temperature combines with habitat destruction to stress both plants and animals. When chronic, these conditions drive animals into cooler areas at higher elevation where forage is less plentiful, thereby reducing reproduction and causing starvation of species (Appendix D).

Chronic temperature elevation also melts perennial glaciers such as the Lyell Glacier that had provided reliable melt water at a seasonal rate. As a direct result, Lyell Lake, historically fed by slow seasonal melt from the Glacier, has been reduced 30% on the western lobe and 70% on the eastern lobe since 1883.[34] Similar elevation in temperature and ecosystem degradation has also occurred throughout the entire central Sierra in California. Whether in the more protected areas of park ecology such as Yosemite or more geographically extended to less protected regions, global warming as a primary cause of environmental degradation has been widely confirmed. A summary of these conclusions indicates that climate change over the past century has adversely affected over 1,500 species of animals in North America. These species so affected by global warming are diverse and "range from molluscs to mammals and from grasses to trees" according to Terry Root, a biologist who also corroborated the 2003

Nature studies (Appendix E).

In particular, the causal nexus has also been forged directly linking regional and global temperature elevation to increased concentrations of carbon dioxide, carbon monoxide and other toxic and/or "greenhouse gases." These atmospheric pollutants are produced from fossil-fueled motorized vehicles, coal and petroleum-fired power plants, industrial discharge and waste from increasing herds and flocks of meat animals. All of these sources directly contaminate aquifers and the methane gas produced add incrementally to atmospheric pollution.

Increasing human pressure from surging populations worldwide preempts habitats of other species, thereby destroying them outright or subjecting them to a slower demise. These swelling human populations in themselves place commensurate demands on available resources to provide only the basic necessities for survival. When large segments of these escalating populations choose to consume in excess of needs, global warming measurably increases, habitat destruction occurs rapidly and widespread environmental erosion takes place. In combination, habitat destruction and global warming factors are lethal and constitute a formula for systemic environmental collapse. The well-documented micro-studies on specific endangered species such as the Quino checkerspot butterfly in Southern California are reliable sentinels for the fate of many macro-populations. These studies quantify the destructive power of global warming and habitat destruction, and provide a scientific basis to predict similar scenarios in much broader populations.[35]

Although the primary causes and their accelerants for species degradation, habitant destruction and global warming have been identified, the specific perpetrators and effective constraints on these agents are still problematic. The actual persons or entities responsible for certain consequences are often opaque and the dynamics involved in even measurable environmental outcomes lack transparency. It may be easy, for example, to cite cases and criteria showing in particular instances a divergence from basic needs to sustain human and animal life. A family of four in the U.S. that lives an indulgent lifestyle, drives 2.6 automobiles, consumes 30% more food calories than needed and uses 400 gallons of water per day, is clearly living beyond its basic needs. When divided into a given population number to reach statistical average consumption, it is no surprise that the metrics of these consumption levels demonstrate that such a lifestyle causes environmental destruction. Meeting basic fuel needs alone for the existing power sources available to this population at the beginning of the 21st century would still consume approximately 16% of the worldwide petroleum supplied by the United States even though its population is less than 5% of the global population.

However, in a consumer and market-based economy operating throughout the 19th and 20th centuries, the United States consumed over 25% of the worldwide petroleum production. Despite these verifiable facts, it is still difficult to clearly identify the moral agents and those ethically responsible for the ongoing destruction of the environment. On an international level, the value networks that create cumulative economic, legal, social and political effects are translated into over-arching conceptual and political policy debates among governmental powers. One such international forum that contemplated governmental value networks on both a conceptual and a policy formation level was the Kyoto Agreement of 1997. The conceptual issues included the reduction of global warming by substantial decreases in toxic and greenhouse gas emissions and limitations on pollution. Despite general conceptual agreement regarding these goals, only 150 countries were signatories with the United States and Australia still refusing to sign the Agreement in 2006. However, it must be noted that China and India, the two most populated countries in the world are exempt from Kyoto provisions as "developing nations." Russia, also has engineered broad, self-serving exceptions under the Kyoto Treaty wherein the provisions grant Russia an exceedingly generous inventory of pollution units that are traded in the lethal and ghoulish commerce of toxic transactions.

Despite the continuing and cumulative weight of evidence showing worldwide erosion of the environment by nearly every measure, even those closely associated with national parks are reluctant to address these urgent issues. The chilling effects of the United States government's positions and policies present under the G.W. Bush administration increasingly inhibit discussion and quash decisive action to abate environmental destruction. This reluctance on both levels of conceptual and policy debate by national park administrators is summarized by a United States National Park official:

> Scientists, naturalists and educators in every park are facing it, and they see its really powerful and inescapable... The biological reality we see around us, and the reality the G.W. Bush administration sees, do not come together very well.[36]

This "reality" is essentially confirmed by the G.W. Bush administration's inalterable agenda in opposing any mandatory regulation to reduce pollution, control greenhouse gases and reduce global warming.

However, within the groups espousing public policy and implementation "realities," there are factions that show conflict with one another. Some of these disagreements are substantive wherein divisive issues involve the fundamental premises of the group's agendas to devise and implement policy regarding land use. Issues of administration for these programs, including scientific studies, compliance or regulatory and legal authority, have been topics of contention. Even within the context of a political administration wherein these substantive issues have been settled by legislative, political or executive fiat, there are serious divisions among the agencies themselves as to implementation, emphasis or nuances of policy programs. Rather than insuring unanimity on every issue for those adhering to the policy and implementation of land use agendas, membership in these groups are fraught with disagreements within a multiplicity of sovereign and sub-sovereign centers of political power.

An example of this type of disagreement within a sovereign administration is found within an agency of the United States government that makes public policy for National Parks' land use. In July 2006, the National Park Service Director, Fran Mainella resigned after five contentious years that were marked by intra-agency strife, sharp criticism from Congress and attacks from groups inside and outside the government. A common denominator substantive issue from these critics was that the National Park Service under Mainella's direction favored using National Parks primarily for recreation rather than conservation. Revisions of management policies during Mainella's tenure had not been completed when her resignation was announced.

As an economic, political and military super-power, the United States has also become the primary sentinel nation for rampant consumerism. A dominant element within the national value network, consumerism creates a mandate to its people that encourages, then coerces them to seek endless wants. Even basic needs become eclipsed not only for the poor but also for the affluent who purchase desired goods and services to the exclusion of the necessary. The cumulative effects of these values and behaviors produce unmistakable environmental deterioration but are subordinate and irrelevant to governmental policymaking at the beginning of the 21st century. As the foremost spokesman for the business-consumerism-political complex, the G. W. Bush administration graphically illustrates the irrelevance of scientific facts to political policies regarding the environment. In a major speech to the Group of Eight Summit attended by eight of the most wealthy and powerful nations in July 2005, President George W. Bush sets forth these flagrant incongruities and contradictions. He stated,

> I will recognize that the surface of the Earth is warmer, and that an increase in greenhouse gases caused by humans is contributing to the problem.[37]

Despite these remarks, toxic and greenhouse gas emissions with residual global warming have continued unabated in the glaring absence of governmental regulations of corporate and consumer pollution (Appendix F). Slogans or clever phrases have been crafted such as "responsible tourism" and "responsible sustainability" that commonly appear in literature associated with environmental issues but they show lack of clarity in actually ascribing responsibility. The Brundtland Report, for example, speaks with amiable fuzziness about organizations, including corporations and governmental agencies making,

> Key national, economic and sectoral [sic] agencies directly responsible for ensuring that their

> policies, programs and budgets support development that is economically and ecologically sustainable. Industry's response to pollution and resource degradation has not and should not be limited to compliance with regulations. It should accept a broad sense of social responsibility and ensure an awareness of environmental considerations at all levels.[38]

The effects and influence of such a political agenda have extended throughout the bureaucracy of the government agencies and superpower sovereigns also give permission to a broad range of decision-makers in the private sectors. Without specific prohibitions in the form of legislation, regulations, public policies and court holdings, the insatiable drive of businesses for profits and power is unleashed. The ancient adage in the Anglo-American legal theory "that whatever is not specifically prohibited is allowed" is taken as a prescription and justification by corporate directors.[39] Armed with this militia and the acquiescence of neutered bureaucratic minions, torrents of consumer products gush forth from industries that, in turn, create cascades of pollutants as by-products of uncontrolled production. This lethal cycle of demand creation and demand fulfillment of consumer wants has inevitably resulted in catastrophic collateral damage to the environment. This colossal machinery, the moral equivalent of vast military machines has also obscured the real moral agents and those ethically responsible for the consequences of individual and collective choices.

Chapter Six

Collective Rule Construction and Personal Ethical Decisions

Tensions regarding decisions made by an individual person and collective rules have been problematic for millennia throughout multiple societies and cultures. These tensions extend widely into various decision-making contexts wherein general rules are established as standards by which individual and collective decisions are made, and by which human actions are judged. Such contexts include legal arenas espousing the "rule of law," religious sects, cults and institutions imposing moral rules, and business, social and political organizations that establish rules of compliance as a function of their policies and procedures. Despite the diversity and ancient lineage of these tensions, the conflicts generated from the interaction between individual choices and rule-governed decisions have remained largely unresolved.

The lack of definitive resolution or even tentative solutions to this continuing tension is, in part, not total dereliction by thinkers responsible for clarifying these issues, but due to the dynamics of these interactions. In the legal context, for example, the "rule of law" admonition has arisen from a combination of legislation and case law created within the maelstrom of economic and political interests. Regardless of whether these interests are reasonable, fair or just, once embodied in authoritative legislation or regulations the rule of law demands obeisance to its prescribed rules. Violation by an individual or collective entity results in the imposition of penalty and punishment by the coercive powers of the sovereign authorities. Such rules are often derived from more complex sets of regulations, including legislation that is embodied in a compendium code. Ancillary sources of rules that impose obligations upon individuals and groups within societies are court opinions and holdings that have accumulated over many centuries to form a loosely organized body of precedents. These precedents become the basis for continuing traditions cited and reinforced through continued acceptance by the courts.[39]

Whatever the legislation or settled case law within a common law tradition, their provisions are considered authoritative only if derived from a more comprehensive source such as a constitution. The constitution that is ratified and accepted within a social and political structure and becomes a common fund of terminology defines the common ground for broad conceptual arguments and for narrowly focused policy debates. From this common denominator for human behavior, minimum standards for individual and collective human behavior are established. It is also by those standards that acceptable or unacceptable behaviors are judged and by which both conceptual and policy-driven debates are decided. Some of these issues are decided by drawing bright lines easily seen by most anyone while others show diminishing degrees of brightness that are fraught with ambiguity and vanish into crepuscular darkness. Whatever the clarity of these laws, both collective groups and individuals are expected to abide by them at their own peril such that "ignorance of the law is no excuse" has been a long-standing legal maxim.

Despite the professed ideals of neutrality lauded in legal literature to be used in constitutional interpretation, elements that lack neutrality have posed chronic threats to the application of these ideals.[40] Included in a litany of threats to the neutral integrity of constitutional interpretations is ethics itself in the form of collective moral pronouncements and individual behaviors claiming to manifest moral rectitude. When imbedded in religious institutions, the pronouncements and behaviors are reinforced by the influence, authority and coercive powers available to those institutions often based on divine right of the leaders and the righteousness of the disciples.

These elements are found in every sovereign nation that is governed by a written constitution and administered according to the rule of law derived from such an organic document. The threats to its even-handed interpretation and adjudication are not only from the righteous rantings of individuals claiming moral authority but also from well-organized coalitions of church organizations that advance agendas based on moral rectitude. These positions extend across the spectrum from the well-organized and financed Christian Right Conservation organization to mainstream churches of every denomination, and to a broad range of charitable organizations directed by the distribution of the wealth mandates or purchased influence.

The chronic problems of constructing coherent and consistent interpretations of a constitution-based rule of law are often compounded by competing voices from these morality-based groups. Rather than offering reinforcement for neutral principles of constitutional interpretation as common grounds for discussion, morality-based agendas often militate against judicial restraint, equitable legislation and unbiased rules of law. The National Christian Foundation founded in 1982, for example, operates as a clearinghouse for funding conservative evangelical agendas that include financing the Alliance Defense Fund and the American Center for Law and Justice and many others that "further the gospel of Christ." Intelligent design, construed as a political action item through the Discovery Institute, is also financed by the NCF. Apart from critiques of the opacity of its operations, self-dealings and political machinations, this group has funded its multiple agendas with nearly $1 billion dollars.

These kinds of organizations have proliferated in societies and cultures for millennia but they are especially prolific within societies that allow diversity of religious cults, organizations and institutions. In contrast to totalitarian political regimes that crush diversity in most every form, a representative democracy supported by a written constitution promotes dissent and diversity. However, in the Western world until the ratification of the *United States Constitution* in 1789, political structures including representative constitutional democracies were inextricably entwined with religion; thereby, the authority of its constitutional laws ultimately was founded on divine authority. Even the *Articles of Confederation* of 1781, the precursor document to the *United States Constitution* contained vestigial references to "the Great Governor of the World" that was implied by the Declaration of Independence allusion that "all men…are endowed by their Creator with certain inalienable rights." Despite this political, theological and legal legacy, the *United States Constitution* and others modeled on it were ratified by the people's representatives as a purified secular basis of authority residing in the people.

Nearly two centuries after the ratification of the *United States Constitution* as a secular foundation of authority, President Ronald Reagan alluded to the Jeffersonian "wall of separation" and the demand for neutrality in the government's discharge of its constitutional obligations:

> …At the same time that our Constitution prohibits state establishment of religion, it protects the free exercise of all religions. And walking this fine line requires government to be strictly neutral.[41]

While purging a document such as the *United States*

Constitution of religious premises would appear easy, since its secular context is devoid of references to the divine, preventing religion and its moral teachings to distort neutral principles of constitutional interpretation has proven to be difficult. Although ostensibly operating outside the secular constitutional text, those exercising religious beliefs and expressing moral agendas obtrude into constitutional interpretations on every level. For judges, legislators and organized groups, the free exercise of religion permits their practices to encroach into the formation and enforcement of collective rules governing human behavior.[42]

Due to this avenue of access to rule construction and enforcement under the color of the free exercise of religion, the fundamental issues found in James Madison's *Memorial and Remonstrance* of the 18th century are still alive in U.S. Supreme Court decisions of the 21st century. Madison's polemic was instrumental in defining the distance between religion and government necessary to insure the proper functions of both. Thomas Jefferson's "wall of separation" speech helped to erect a graphic picture of Madison's impassioned arguments that sought to limit the distorting effects on both sides of the "wall." Despite the literal cleansing of the *United States Constitution*, leaving it devoid of divine references and the construction of solid rhetorical walls reinforced by lucid Jeffersonian prose, the *aretean* ideal continued to be breached by a multitude of Supreme Court decisions.

Because of the porous properties of this wall, separation of church and state has continued to be honored sometimes in the breach rather than in the observance of this disjunction. In this way, Justice Antonin Scalia was able to base his dissent in *McCreary County v. the American Civil Liberties Union of Kentucky*, decided in 2005 on religious and moral beliefs that are brought into the decision-making process.[43] Such beliefs enable an U. S. Supreme Court justice to erode the neutral principles of constitutional interpretations and introduce elements that militate against Article 6 and the First Amendment establishment clause of the *United States Constitution.*

Due to the malleability and porosity of organic documents such as a constitution, interpretations of its fundamental premises are varied and can be disputed. However, the individual citizen is faced with the responsibility and duty to obey a constitution and its moral and legal imperatives that are fraught with ambiguity, contradictions and distortions caused by elements that change the neutrality of interpretation. These same issues arise from a wide range of other organic documents found at the basis of many institutions, including churches that derive their ethical teaching from those foundational premises. One centralized church organization that offers a representative example is the Catholic Church, which has predicated its moral theological teaching on the writings of Thomas Aquinas. While articles of faith have been propounded as dogmas derived from Biblical exegesis, Thomistic authority has primarily supported moral theology as a reasoned body of teachings. Since the condemnation of his writings was revoked 50 years after his death in 1274 C.E., Thomistic writings have constituted the principle organic theological and ethical teaching authority of that Church.

From this fund of information, based on the extensive writings of one person, has developed a system of ethics based on essences, general norms and natural law that has been advanced as the official teachings of the Catholic Church for nearly 700 years. Similar to an organic constitutional document, ethical conclusions are deduced from premises that become general norms, rules and prescriptions by which human behaviors are judged. Both an organic constitution and a systematic compilation of homogenous ethical premises construct collective rules that bind particular human actions and create imperatives for personal ethical decisions.

However, the Thomistic system has been under attack from various assailants for centuries, including profound critiques from within the Church institution itself. During the last half of the 20th century, Karl Rahner, a Catholic Church theologian, launched a broadly based assault on the validity of general ethical norms or collective rules that purport to govern human ethical choices. While broad in scope and fundamental in its outcome, Rahner focuses on the very specific uniqueness of each person as a free ethical agent. Owing primarily to the singularity of each person, general prescriptions derived from equally general rules cannot generate adequate authority or direction for individual ethical decisions. In summarizing this position, Rahner states in a seminal article published in 1950 C.E.:

> Thus, there is a realm of individual morality and religion a region of moral and religious obligation and duty which, without opposing the general law of morality, lies decidedly outside this pale and cannot be prescribed within their general formulations.[44]

Due to the irreducible uniqueness of each person, according to Rahner, efforts to establish general rules that collectively bind an individual's ethical decisions and moral behavior are futile. Appeals to essences, general norms or natural laws for collective rules to prescribe individual ethical choices will inevitably fail.

Rahner recognizes that there has been "a strong proclivity for rationalism and idealism" that reinforces the deduction of collective rules from *a priori* premises. Despite this proclivity, the powerful arguments advanced by the advocates of phenomenology and existentialism provide a cogent and compelling rejoinder to the venerable Thomistic tradition. Although admittedly comforting to deduce ethical prescriptions from general premises, this method creates a widespread illusion of certainty for the individual ethical agent. As a deeply entrenched and venerable tradition, a deductive system of *a priori* premises such as Thomistic Moral Theology enjoys pride of place because of its long and venerable tradition

(Appendix G).

Once the "ontological uniqueness" of each individual is fully understood and embraced, the fragility of any system of collective rules and general prescriptions becomes increasingly apparent. As ontologically unique, the individual person must then be seen as irreducibly singular, subject only to its own singular source of ethical obligation found, according to Rahner, in each person's conscience. The *gravitas* of this indefeasible singularity generates the ethical imperatives from a source of moral authority that is located within a well-formed conscience. Rahner was aware of the Freudian definition of conscience as the internalization of external authority figures and their directives. However, Rahner's critique departed from Freud's pragmatically useful but inadequate parameters that suggested a too simplistic model of conscience wherein its functions were limited to mirror imaging from selected authority images randomly occurring in an individual's life experiences.

On such a mechanized model, a person's conscience would be essentially a function of adroit manipulation, random selection or image control within a society. Such a person would develop, according to the observation of Plato, as one becoming a more acceptable citizen within a *polis* but one becoming a worse human being. Every society could use its own set of metrics to measure the effectiveness of its own particular methods of engineering or forcing behaviors desired of its citizens. Socrates was executed in 399 B.C.E. by the workings of these metrics, as were tens of millions who have opposed a wide range of external authority figures on the grounds of conscience.[45]

Rather than foisting general prescriptions for ethical choices on a person that are deduced from collective rules or general norms, Rahner advocates the operation of an individual internally-formed conscience that is apart from abstract essences or general rules. Similarly, a Freudian depiction of conscience engineered by manipulation, fear or force leads to the illusory creation of right based on might. Any elements within the Thomistic moral theological system that may be salvaged, according to Rahner, are derived from *a posteriori* induction wherein workable, coherent and authentic general ethical propositions are built up from instances based on individual moral deliberations, virtue and decisions. Acceptance of an inductive *a posteriori* methodology recognizes the significance and indefeasibility of the individual acting on the promptings of that person's internal conscience. The formation of such a conscience is not guided by imperatives derived from social manipulation, political engineering, coercive power of the sovereign, the imposition of general norms, or from putative moral essences.

These issues are central to the continuing debates between the wilderness conceptualists and the policy and procedures-oriented disputants who focus on individual actions rather than seeking enlightenment from abstract essences that emerge from idealized wilderness concepts. Regardless of the content found within an organic document such as a constitution or within a compilation of homogeneous writings, deriving defensible imperatives from those sources remains problematic. In that these sources of authority are used as the basis for collective rules that govern individual actions, the conflicts that result are common to constitutional interpretation, a body of institutional teachings or environmental disputes. As a systematic body of teachings, the wilderness concept has been presented in idealized metaphysical terms that convey a vision of nature as undefiled, indefeasible and untrammeled. In this form, the wilderness concept position is closely analogous to the systematic writings of Thomas Aquinas critiqued by Karl Rahner nearly 650 years after those writings became the official teachings of a major ecclesiastical institution. As *a priori*, both establish conceptual visions from which natural laws, general norms and collective rules are derived.

When an idealized wilderness vision becomes the foundation for wilderness advocacy groups such as the National Wildlife Federation, the premises of these groups are formed into organic documents that become their working constitutions. Once the controlling terms and conditions of their constitutions are put in place, prescribed actions are deduced from these terms and conditions. The functional equivalent of terms and conditions, such as the right to pursue life, liberty and (property) happiness, are the equally indefeasible rights inherent in the unmodified, unchanged or untrammeled wilderness ideal. As an ideal concept and vision predicated on *a priori* premises, purist wilderness advocates are unable to establish détente with policy and mixed-use disputants without compromising their foundational tenents. To do so would be the functional and moral equivalent of a Thomist sanctioning actions that are intrinsically wrong and that violate ethical rules based on *a priori* essences, the natural law and general norms.

Wanton slaughter of innocent children, for example would be deemed a violation of Thomistic ethical norms in any time, place, manner or circumstances, despite the fact that such behavior has been widely practiced and even sanctioned in many cultures by statute. Under the Justinian Code recognized as authoritative law within the Roman Empire, "exposure" was sanctioned whereby parents of an unwanted child were allowed to cast the child into a garbage dump thereby exposing it to nearly certain death. The only requirement imposed by the Code on the parents in this regard was the obligation to notify three neighbors.[46]

Clashes between combatants in quarrels that involve immutable and inflexible premises have frequently been settled not by negotiation or reconciliation strategies but by force. While many historical conflicts among conceptualists have originated on lofty planes by pious proponents against perceived adversaries, the conclusions have not remained on the high ground of essences or purified norms. Whether the conceptual organic document was a *Magna Carta, Declaration of Independence, Declaration of Liberty, Articles of Confederation* or a ratified Constitution, the collective rules, rights or norms were usually sustained by the advance of troops. For ethical teachings adopted as authoritative by ecclesiastical institutions, variations of the Inquisition, Crusades, Jihad and Reformations, or pogroms of generic "ethnic cleansing" or genocide were undertaken to enforce prescribed moral behaviors and actions. It remains to be seen whether those embracing the wilderness concept will escalate their campaigns by means of covert sabotage strategies or gain access to power by political and economic means; or modify their rhetoric and engage in compromises with those making public policy.*

*In a 2007 case involving attacks by radical environmentalists against "mixed use" land development, the Federal prosecutors in Oregon were considering bringing criminal charges under the U.S. Terrorist Statutes. It had been alleged that eco-terrorism in this case involved arson, including incendiary devices to destroy property with foreseeable harm to persons. These elements present in this case make it plausible that "eco-terrorism" is tantamount to terrorists' attacks against a sovereign entity.

Chapter Seven

Sovereign States, Nations and Their Environmental Authority

The challenges facing environmental management and the ethical issues involved are compounded within the context of the sovereign state. Since the sovereign state model in any of its forms uses political power to implement its agendas, fundamental premises of this model become the threshold concerns to justify the exercise of these powers. The history of these justifications includes claims of divine right from which sovereigns derive their powers and by which their subjects are ruled. As an ultimate source of power, divine right allows the sovereign unbridled license to rule the sovereign state without qualification or condition. Whether this model of power is structured as a dictatorship, a theocracy, an aristocracy, an oligarchy or a constitutional republic, sovereign power is absolute. As a theocracy, the sovereign state is an expression of ecclesiastical institutions wherein there is no separation of church and state; a sovereign pronouncement in a theocracy is tantamount to a divine directive, which is unassailable and unalterable by human intervention.

The theocratic expression of sovereignty arose early in ancient cultures in both hemispheres of the world.

Egyptian pharaohs, as theocratic sovereigns, had their Oriental counterparts and pre-Columbian sovereigns in Eurasia practiced theocratic sovereignty in ways similar to the Incas and the Aztecs in the western hemisphere. Political power arrogated by the theocratic sovereign by virtue of divine right precluded the exercise of any significant power by the subjects. Sovereignty founded on an aristocratic model claims its supreme powers are vested in those who enjoy pride of place in the political hierarchy because of birth or appointment. Aristocracy as a social stratum has been formed in almost all societies and perpetuates its standing by strategic marriages within that privileged class. Once seized upon as an entitlement, aristocratic rule has also raised itself to the august levels of sovereignty claimed as a theocratic divine right.[47] Similarly, oligarchs made up of an elite few who claim special talents, appropriate sovereign powers as natural rights. These natural rights of oligarchs are also considered self-evident, inalienable and inviolable as the natural rights underlying a wide range of constitutional republics and democracies.

Despite the particular differences among the varieties of sovereignties and their claimed bases of power, all sovereign powers have expressed themselves in city-states, nation-states or larger entities defined by the sovereign. Central to these political structures is the exercise of power over the subjects wherein authority figures appointed by the sovereign act as power brokers and administrators. When these sovereign powers are embodied in a formal political structure such as a nation-state, the subjects, whether citizens, aliens or slaves, are subject to the coercive enforcement power of the state. Within the coercive confines of these concentrations of power, the subjects of the sovereign are allowed to live at the sufferance of those in power. Here, too, a sovereign's subjects are faced with myriad, daily decisions to be made, many of which involve some aspect of their environment.

The operations of sovereigns over the millennia have shown a wide range of behaviors but have also demonstrated common threads of similarity. Reasons for accepting sovereignty by subjects have been multiple although patterns of similarities can also be recognized. Rule by conquest, inheritance or appointment has been a common justification for sovereignty, although other foundations for sovereign power include the reasonableness of Plato's philosopher king and Thomas Hobbes' Leviathan to which the subjects freely give their consent. For the latter, the Leviathan sovereign liberates the subjects from lives that are "mean, nasty, brutish, solitary and short" in exchange for which the subjects acquiesce to authoritarian rule for "gain, safety and reputation." Plato's benign sovereign represents the paragon of reason that guides and governs a diverse population comprised of "gold, silver and bronze" subjects.

These diverse models of sovereignty and their background origins are relevant to the manner and methods used in environmental management laws and ethics. For example, multiple sovereign nations that have become industrialized nation-states over the past 250 years have implemented policies and procedures that were favorable to industry and the interests of the industrialists. Deference was shown to industry and its agents despite detrimental consequences for the citizens, animals, natural resources and the environment. The profound social and economic consequences not remedied by the sovereigns of industrialized nations were the subject of critiques, parodies and satire by writers, including Jonathan Swift, at the beginning of the industrial revolution. Already by the 1729 publication of Swift's *A Modest Proposal,* the sovereign's governance had failed to provide food, water, shelter and healthcare, and curb pollution or control overpopulation. In England, Hobbes' sovereign Leviathan ruled but had grossly breached its social contract with its citizens; with exceptions for a few aristocrats, contractual promises of "gain, safety and reputation" had been flagrantly broken. In the face of such sovereign dereliction of duty, Swift's parody suggested that small children be eaten to ameliorate woeful social conditions, including both the extreme shortages of food and uncontrolled population.

Similarly, the forms of sovereignty and their governmental agendas affect programs, priorities and policies that directly affect specific environmental issues. Mounting evidence indicates that these sovereign governance priorities and policies have been "captured" by the self-interested industrialists to the detriment of all other interests. Since Swift's early 18th century parody, other luminaries such as Charles Dickens and Karl Marx have attacked the specifics of social and environmental degradation in the 19th century. A common theme of these critiques points to the common-denominator issues that involve the betrayal of the citizens by the sovereign, the destruction of the environment and the eclipse of interests other than industrial and capitalistic interests. Revenues generated by unchecked industrial commerce have swelled the coffers of the sovereigns but at the expense of citizens, animals and their environments.

These tensions and unresolved conflicts continue in contemporary societies as the sovereign nations govern their citizens and control wildlife and their environment with powerful industrial interests inextricably enmeshed with sovereign decisions. The Bern Convention signed in 1970 was a 20th century effort to interject countervailing measures to "protect and restore the biological integrity of their native animal communities." Prior to this international convention, each sovereign signatory nation exercised its own sovereign powers to manage its wildlife and its eco-system environment according to its own prerogatives. The cumulative results of these former programs ordered and administered by those nations included a continuing destruction of most all species of wildlife and its habitats that lead to the extinction or near extinction of many forms of life. In some cases, the erosion was precipitous in that outright shooting, trapping or poisoning of animals sharply decreased wildlife populations. Equally effective in exterminating large populations of wildlife species was the systematic dismantling of wildlife habitats by uncontrolled land use development, rapacious deforestation by timber companies and direct and indirect poisoning of food and water sources. The saturation of DDT, for example from 1945 until its ban a quarter of a century later, caused widespread destruction of both plant life as a food source and animals, especially bird populations. This international ban on DDT is still breached by China and India where it is produced and used widely.

Management of North American and European wolves illustrates the policies and priorities of sovereigns since the advent of the Industrial Revolution. The wolf, a representative victim of this habitat and wildlife destruction, is an animal common to all Bern Convention signatory nations and had flourished in these jurisdictions for millennia. A natural balance of the wolf within its habitat was maintained by pack adaptation. Its pattern of predation evolved such that weak, sick and old prey was culled out by wolves, thereby establishing a natural cycle of life and death. Wolf populations decreased under these habitat conditions when herd count of their prey diminished and increased when prey was plentiful; however, wolf habitats worldwide were systematically modified, then broken by human encroachment. Although each of the industrialized countries had directly affected this ancient wildlife pattern on their own timetable using a variety of methods, the impact on wildlife, including wolves, was the same. Despite the superficial differences of each of these sovereign nations, the policies initiated by each of them for wildlife management, including wolf packs by industrialized nations, were nearly identical. For all wolf populations, they were methodically decimated such that the total numbers in each of these nations declined to less than 1,000 thereby causing wolves to become a highly endangered species in all the sovereign nations (Appendix H).

While the near extinction of all wolf packs in the industrialized nations was allowed by the decisions and policies of individual sovereign nations, remedies, even partial ones, were recognized to be implemented only by concerted efforts such as the Bern Convention. Pursuant to the terms of such a treaty, each nation is limited in its exercise of unlimited sovereign authority by the countervailing powers of the signatory countries. Even though the signatory nations to an international treaty such as the Bern Convention assent to its terms and conditions, a breach of an international environmental accord by any signatory nation is usually without penalty. Similarly, withdrawal from an international or intranational treaty by a nation-state may result in a formal sanction but is devoid of real punishment unless actionable under the jurisdiction of an international tribunal or enforced by trade restrictions imposed on hostile nations.

Pursuant to these prerogatives and powers of

nation-states, the United States, for example, had a long intranational history of unilateral breaches of treaties with Native American tribes wherein treaty terms and conditions were unilaterally rescinded or breached at will without penalty. Native American "sovereign nation" status of indigenous tribes eventually emerged from the Indian reservation system. Also, the "combatant" status that gave relief from criminal prosecution was granted to these peoples during colonization of the Western Hemisphere. As such, the Native American and Arctic People tribes retain a tenuous sovereign nation status at the sufferance of the U. S. Government, primarily as symbolic largesse, a forum for legislation, standing for litigation and for economic development such as casino businesses that are exempt from the sub-sovereign authority of the several states.

The impasse experienced among sovereign entities, including city-states, nation-states or countries when an agreement or treaty is no longer acceptable to a signatory, illustrates the self-limitations of a sovereign model. Whenever one or more of the sovereign parties to the agreement reject that agreement, such a unilateral rescission is usually done with impunity. This has been the commonly occurring result of most all intranational and international environmental agreements such as Kyoto Treaty or the Bern Convention. Even though a specific environmental topic has profound global implications and is based on extensive scientific findings, an intranational initiative may be aborted at will or its own findings largely ignored by other sovereign nations and the sponsoring nation itself. The rejection of the Kyoto Treaty by the United States under the George W. Bush administration was reinforced by the Republican-controlled Congress in its 2005 refusal to pass the McCain-Lieberman anti-pollution bill, sometimes referred to as "Kyoto Lite."

The United States Global Change Research Program initiated in 1989 is an example of this type of intranational dissonance. Its policy statement was predicated on the role of scientific research in forming environmental policies regarding climate change worldwide.

> The U.S. Global Change Research Program was conceived and developed to be policy-relevant and hence to support the needs of the United States and other nations by addressing significant uncertainties in knowledge concerning natural and human induced changes in the Earth's environment. The U.S.G.C.R.P. is designed to produce a predictive understanding of the Earth system to support national and international policymaking activities across a broad spectrum of environmental issues.[48]

Since that statement was made at the outset to establish the goals of the Program, $25 billion has been spent on research directed to those objectives. The earlier predicted increase in global warming during the 21st century was increased from a 1.4 degrees Fahrenheit prediction in 1990 to a 5.4-degree increase based on the Program's findings. Other than concluding in general terms that average ambient and atmospheric temperatures will rise, thereby causing global warming, and that human behaviors are the proximate causes, little else has been accomplished. Although the program continues to cost U.S. taxpayers $1.8 billion per year, its contributions to the environmental policies and priorities of the sovereign nations have been negligible. Even though general and specific conclusions of this enormously expensive program have been published, the sponsoring nation itself has been unresponsive in terms of policy formation based on this extensive scientific research. Because of the gross indifference of the United States to its own sponsored climate research directly impacting environmental policies formation, it is not surprising that other sovereigns turn an equally deaf ear to scientific models useful to making environmental policies for their own nations and for international accords, treaties or conventions.[49]

The power arrogated and available to sovereigns and their fiefdoms, structured in multiple ways, allows the exercise of their coercive authorities to benefit or harm the global environment. Within the confines of a sovereign jurisdiction, both the vision of eco-systems as idealized concepts and the formation and/or implementation of environmental policy are at the sufferance of the sovereigns and their non-binding environmental resolutions. Voices of protest and dissent within these fiefdoms are subject to the sovereigns' ears that may be muffled or entirely deaf. Influence peddling, self-serving legislation, campaign contributions, court-packing or a wide range of corrupt practices distort or control sovereign choices.

These same forces, along with economic and military power issues, directly affect policies and decisions made by sovereign nations regarding international environmental treaties. Despite the pretense of international diplomacy to assuage antagonistic interest groups, the more powerful sovereigns meet at will to allocate pollution limits measured in exchangeable global warming units. This agenda of *Real politik* continues with its damaging consequences for eco-systems worldwide, despite copious, convincing and compelling scientific evidence to the contrary (Appendix I).

CHAPTER EIGHT

MODELS OF SCIENCE AND SCIENTIFIC RESEARCH AS ENVIRONMENTAL ETHICS GUIDELINES

The model of sovereignty with its various manifestations and morphologies has been inextricably linked to intranational and international policy formation and execution. Sovereigns have taken the initiative to control and dominate these arenas by virtue of claimed positions within their controlled and circumscribed jurisdictions. External to these political and legal parameters, sovereigns exercise global prerogatives by strategic alliances, economic-based diplomacy or military force. However, on both intranational and international fronts, the sovereign model reaches its outer limits for even-handed or rational decision-making, including environmental conceptualization and policy formation. These limits have become, in fact, tangible barriers to dialogue, problem solving and management of the environment, such that sovereignty and its expression in nation states have failed to provide either an effective or rational model for environmental concepts, policies or ethical guidelines.

Once confronting the weaknesses of sovereignty and recognizing sovereign immunity to international or intranational policies and treaties deemed unacceptable, the role of science and scientific research within environmental debates needs to be considered. While both may be analyzed in the abstract, their relationship to the debate needs also to be evaluated in terms of their impact on choices and decisions in a world governed by sovereignties. The products and by-products of science, for example, are expanded and/or contracted in differing ways within sovereignties, some of which systematically suppress selected scientific innovations while others are allowed or promoted by government funding. Similarly, scientific research and its applications are also fostered, inhibited or quashed not due to a lack of scientific merit or rigor but because of arbitrary sovereign declaration. In some significant instances, continuance of redundant or even discredited scientific research is funded by generous government budgets in order to postpone political decisions and/or action. It is always convenient or expedient to demand additional scientific data prior to a decision as articulated in *Frontiers of Illusion* by Daniel Sarewitz.[50]

One of the central premises of Daniel Sarewitz's book and of others addressing the role of science in policy decisions is that scientific findings are uniquely suited to settle conceptual and policy debates. Rather than contributing to partisan arguments and prolonging political quarrels, science is viewed as the final arbiter and the means to arrive at undisputed answers. Part of this appeal is rooted in the belief that scientific objectivity is capable of replacing the subjectivity of economic, political and religious interests. Subjectivity is thought to be fraught with uncertainty and purging it from serious discussions would substitute the certainty of science for endless ambiguities and flagrant falsehoods advanced by subjective and self-serving partisans. The ideology of such a scientific quest is an expression of rationalistic models proposed by Plato, reduced to a methodology by Descartes and further refined by those who have articulated their confidence in the existence of an objective world. Such an objective world is knowable by the human intellect according to this ideology and can be expressed in propositions verified and/or replicated by empirical means. A reality characterized by Immanuel Kant as the *Ding an sich*, the thing as it is in itself, is an expression of the same ideology that urged Albert Einstein to explain the world as a reality free from the caprice of playing dice.

However appealing this ideology is or however enticing this belief is, the model of scientific objectivity itself has been under attack since its original formulations and has been the object of systematic critiques even in its highly sophisticated forms. The association of objective science with subjective policy formation has been itself a forum for continuing argument; and attempts to settle partisan debates by invoking science have reawakened a broad range of issues involving the capabilities of science to provide definitive answers to devisive questions including those affecting the environment.

When called upon to provide dispositive proofs to validate either a broad conceptual or policy issue relating to the environment, scientific findings themselves are used or ignored according to the agendas of the disputants. The relevance of scientific facts or data themselves to environmental discussions is being called into question, as is the overlapping relationship of science to conceptualization or to policy resolutions. Beyond a sovereign's rejection of any science by *fiat* as the exercise of sovereign prerogatives, there has also been resistance to scientific findings based on other premises. These premises go beyond speculative metaphysical and epistemological issues that call into question the real objectivity and credibility of scientific propositions. While such critiques continue and are constructive on various levels of understanding, science and its direct applications in policy formation have been held up to another criticism: science or technology innovations are inadequate to justify or compel moral policy formation or to frame ethical conceptual ideals on their own.

A useful example to illustrate this impasse can be found in the scientific and technology research funded by the National Institute of Health (NIH) directed toward medical practice and healthcare improvements. Similar to the U.S. Global Change Research Program (USGCRP), the NIH is funded by the government in the amount of $20 billion per year that has included several billions of dollars each year for stem cell research. Within the context of biomedical research the general and specific shortcomings of science and technology for policy or concept formation can be seen. Arguably, all of biomedical research and applications have far-reaching implications beyond science and technology themselves. Ethical, social, religious, legal and economic issues are inextricably enmeshed with NIH projects, including fundamental bioethical concerns arising from stem cell research at the outset. At this threshold level, the science and technology involved with stem cell research have been under attack as being inherently unethical because of the outcomes and the technology processes used. It is argued that the derivation of stem cells from aborted fetuses as primary research material is inherently unethical, whether from intended abortions or from unintended fetal demise.[51]

Moreover, direct outcomes from many NIH research projects including stem cell research have profound effects on humans, animals and their environment. Any bio-engineered plant, animal or human resulting from NIH research projects will necessarily involve choices made on non-scientific and non-technological premises. If the choices allow certain forms of life to be extended, certain environmental consequences must be expected, including the proliferation of certain species, the multiplication of selected breeds, and individuals with selected traits within those species. Profound questions regarding the quantity of certain forms of life that result from stem cell research, along with lives extended by other medical interventions are not readily answered solely by science or its methods. Biological science and its frontiers advanced by NIH funding have produced medical devices, drugs and intervention therapies that directly decrease infant mortality, prevent life-threatening diseases and prolong the end stages of life. In consort with ancillary technologies, medical and healthcare research have contributed to average human life expectancy that has increased over 25% during the 20th century.

Beyond these direct consequences of scientific research and technological innovation, equally problematic are issues arising from NIH-funded projects that affect organ replacement, the Human Genome Project, genetic engineering and nanotechnology. Each of these areas is fraught with a broad range of implications that is not amenable to scientific or technological solutions alone. Scientific or technology innovations that would regenerate diseased organ tissue or abate rejection of transplanted organs using embryonic stem cells would appear benign, but they actually produce significant social impacts as consequences. In addition to the increased population numbers, the costs for treatment would amplify socio-economic inequality by allowing the wealthy first access to the innovative medical therapy. These concerns would be in addition to the threshold ethical issues regarding the acquisition and use of these stem cells or genetic engineering in the first place.

Due to these real concerns arising from scientific research and technological innovation, including healthcare and medicine, the Human Genome Project has earmarked up to 5% of its budget for the study of ethical, legal and social implications of research. In addition, other research agencies have proposed to establish Ethical, Legal and Social Implications (ELSI) programs that would explore and evaluate the implications of their research projects. The Department of Energy, for example, is moving toward the creation of an ELSI program specifically to be integrated into research in nanotechnology. By 1997, the National Science Foundation had established criteria for funding scientific and technology projects that required an ancillary review process directed at the broader impacts on society and the environment (Appendix J).[52]

The view that science needs to be subject to ancillary review and scrutiny is due not necessarily to the view that scientific method is unreliable, but that it may be considered too reliable. A very high degree of acceptance for scientific research raises its conclusions to the unassailable position of infallibility. Although "falsifiability", to show counter evidence has been fundamental to scientific progress, the continued absence of rigorous scrutiny has allowed certain scientific theories and propositions to become "normal science."[53] Practicing normal science creates predictable results until anomalies accumulate to force a paradigm shift, thereby bringing about a scientific revolution. Since revolutions are infrequent, normal science allows long periods of equilibrium in the scientific community that is punctuated only when these radical paradigm shifts occur. These long durations of stasis are the more general accounts of science that were characterized by Stephen Jay Gould as "punctuated equilibrium" which he observed in the evolution of life forms.

During the hegemony of normal science within various scientific areas, the dominant paradigms allow those within those fiefdoms a homogenized view of reality. Outside of these circumscribed parameters there is little credence given, few new scientific agendas launched and no funding advanced for either the scientists themselves or for those developing ancillary technology. Kuhn's own statement captures the force and profound effects of these oppressive constraints on new scientific discovery:

> Closely examined, whether historically or in the contemporary laboratory, (normal science) seems an attempt to force nature into the preformed and relatively inflexible box that the paradigm implied. No part of the aim of normal science is to call for the new sorts of phenomena; indeed those that will not fit into the box are not seen at all. Nor do scientists aim to invent new theories, and they are often intolerant of the theories invented by others. Instead, normal scientific research is directed to the articulation of those phenomena and theories that the paradigm already supplies.[54]

As a general observation, Kuhn's normal scientific way of knowing affects everyone's view of the world and defines the general ways in which this world works. Humans and animals perceive and understand according to learned ways that define reality and overlook or reject even anomalies that are not supposed to be present. Any anomalies that could be used to falsify certain aspects of scientific research are dismissed outright or are managed by the construction of *ad hoc* hypotheses to salvage the normal science research and confirm the dominant paradigm. The history of science abounds with large and small examples that serve to support Kuhn's analysis and show that ambiguities, anomalies and outright contradictions have been ignored in order to preserve the "normal science" areas of science. A geocentric view of the cosmos persisted for millennia despite the glaring presence of ambiguities, anomalies and contradictions until their cumulative weight and the boldness of Nicholas Copernicus overturned the false but normal science belief and the paradigms supporting them. Destruction of these false and contrived paradigms results in a scientific revolution that brings about cataclysmic change and profound reconstruction required to erect new systems to supplant those discredited.

Although these transformations have occurred in every scientific specialty area over the millennia, actual scientific revolutions are infrequent and sporadic. Their lack of frequency throughout human history is due to the many forces that resist change. These forces have provided long durations wherein orthodox but false science, or respectable yet unproductive scientific research, continues to be supported. Funding from multiple sources is channeled into scientific research projects that are deemed to be reliable, predictable, orthodox and representative of normal science. Predictable results with a high degree of probability and reliability are generated from these projects, many of which reflect the agenda promulgated by the source of funding. Scientific research that is driven by political, economic or social agendas is in sharp contrast to a traditionally conveyed view that science produces objective results that are verifiable, falsifiable and true in all places and times. Agenda-driven research is also in sharp contrast to normal science that is comprised of paradigms that may or may not be true or verifiable at any given time.[55]

During these protracted periods of time, normal science and its supporting paradigms reign supreme and science becomes both increasingly dogmatic yet most suspect. The dogmatism allows its pronouncements to wield great authority from positions of power and to dissolve competing viewpoints. As specific areas of dogmatism become increasingly entrenched, the information results become commensurately homogenized and banal. An illusion of certainty is created during these times of quiescent creativity wherein the informational content of science becomes less fertile although more pulverized. Verification of this highly fractured informational content tyrannized by normal science is similar to confirming a newspaper article by reading another copy of the same paper.

Under these conditions, scientific findings are scripted, thereby diminishing their value as supportive or dispositive of multiple significant issues. These characteristics of science impact the credibility of environmental advocates and those with opposing environmental views. Those seeing an advantage in postponing environmental policy decisions may insist on discrediting the findings of normal science reports that run counter to their own agendas. Others may demand decisive environmental policy reform action within a narrow time-frame by arguing that the "facts are in " which facts convincingly support their agendas. A small but passionate few may initiate necessary paradigm shifts that overturn normal science that dictates consensus at certain centers of environmental policymaking.

Apart from these caveats and limitations, scientific research and its conclusions even at their best do not fully answer the central concerns of environmental ethics. Despite the benefits of objectivity, verifiability, falsifiability and reliability, the scientific method only generates descriptive propositions rather than prescriptive statements. Regardless of whether the science practiced is based on antiquated paradigms that have not shifted or are based on substituted paradigms, the work product of these processes is still descriptive whether true, false or ambiguous, of how the world works. Although scientific findings may provide an empirical basis for descriptions about the behaviors of humans or animals in various habitats, these descriptive conclusions do not in themselves give ethical authority for how these subjects ought to behave.

Over the centuries efforts have been made to ignore or minimize the categorical differences between descriptive propositions and prescriptive ethical statements. Natural law, as the rational basis of ethical systems articulated by Aristotle in the 4th century B.C.E., de-emphasizes the fundamental differences and derives ethical prescriptions from descriptions of nature. In that science contributes to these descriptions of nature, scientific research and its conclusions are considered directly relevant to natural law ethics. Scientific conclusions that can form a general view of the universe along with the earth and its ecology have been incorporated into the *gestalt* worldview of the natural law ethicists. The relationship among the inhabitants shaped by a common worldview and their environment is defined in theory and in the practical application of natural law. Such a worldview allows for the deduction of general natural laws that are further reduced to more specific ethical rules that guide human behaviors within a habitat or a given eco-system.[56]

However, the credibility of this venerable and broadly based natural law tradition is eroded when bogus "science" is used to form its foundations. Even if such an effort to derive prescriptive ethical propositions from descriptive statements is considered plausible, false descriptive statements including those supplied by false scientific research destroy any prescriptive propositions. As a result, ethical statements, including those used in the ongoing environmental debates, loose their moral authority. It remains to be seen whether or not arbiters may emerge that prove capable of separating authentic science from the counterfeit pseudo-science that severely distorts scientific research, theories, paradigms and applications to environmental issues at any level of the discussions. In the absence of such a competent and credible arbiter that separates fact from fiction and maintains its objectivity without self-serving agendas, rational environmental ethics will be replaced by other agendas driven by alternative forces including those controlled by large corporations.

Chapter nine

Business Models, Agendas and Their Affects on Environmental Ethics

The promise of scientific research and scientific conclusions to provide a credible foundation for environmental ethics has been kept only partially at the dawn of the 21st century. Although powerful in its methods that reach conclusions by induction from empirical data, along with prediction and replication of experiments, science at its best remains limited to earthbound descriptive statements. These statements have been incorporated into ethical prescriptions such as those proposed by the natural law proponents and they have been extended into the environmental ethics debates by derivation. Credibility, reliability and predictability as characteristics of statements relevant to environmental debates are derived from their scientific association. While certain explicit and implicit assumptions are needed to infer an ethical prescription from a descriptive proposition, the factual foundations can be firmly built from both normal science and paradigm-modified science.

Despite the continuing advances of science to elucidate and confirm relevant facts regarding the environment, the promise and potential have been eroded, distorted and eclipsed by antagonistic groups with fundamentally divergent goals. One such group that stands in opposition to authentic science consists of corporations that share an agenda that is contrary to science and hostile to its findings. Membership in these increasingly consolidated groups has expanded to include small, medium and large business corporations that share common goals. These goals are both proactive and reactive such that proactive initiatives are taken to maximize their profits by most any means including financial support of self-serving "research" that supports their own objectives. As reactive, members share knee-jerk reactions along with more deliberate campaigns to discredit scientific conclusions that have been validated by peer reviews. Swelling membership numbers have translated to increasing resources being allocated to their agendas and their influence expressed by mounting campaign contributions to compliant politicians or religious organizations.

A graphic and instructive example of this agenda perpetrated by a multinational corporation is found in the business activities of Exxon Mobil Petroleum Corporation. This corporation paid its chief executive officer (CEO), $688 million in 2005 as a retirement bonus for his leadership of a company that has been a leader in environmental pollution for decades. When confronted in the early 1980's with early scientific studies substantiating ozone depletion, greenhouse gases, pollution and global warming, industrial corporations including large oil, automotive, chemical and mining businesses formed the Global Climate Coalition specifically to oppose any restrictions on pollution of any kind. This opposition included influence peddling by purchasing access to politicians through large campaign contributions and paid lobbyists. This corporate strategy has proven to be favorable for the Global Climate Coalition members from the early 1980's and it continues into the 21st century. Its effectiveness is quantified by the relevant measure showing that almost no restrictions have been placed on corporate pollution and its deleterious environmental consequences.

As further reinforcement of the blatant purchase of politicians' votes favorable to the Coalition's business interests, the Coalition led by Exxon Mobil launched a well-financed program to discredit any scientific research that was not amenable to its own agenda. In an article written by Paul Krugman of the New York Times News Service an internal memorandum to participants at a 1998 American Petroleum Institute meeting recommended a systematic program to discredit all scientific studies that show increasing pollution, ozone depletion, greenhouse gases or global warming.[57] As a further extension of this negative strategy, the Coalition of Petroleum Corporations initiated an ongoing collusion to oppose any efforts to reduce toxic emissions. This agenda of obstructionism was methodically pursued despite mounting and credible evidence showing the disastrous consequences of this course of action.

In addition to these well-financed programs to

discredit and obstruct, the Coalition and Exxon Mobil independently have undertaken the lavish funding of their own propaganda mills. These public relations arms of the petroleum corporations essentially are directed toward generating massive amounts of fictoids to support self-fulfilling prophecies. The pre-programmed "crystal ball" of these minions is not the result of dispassionate scientific research but is instead the manufactured justifications of conclusions already made. Rather than putting relevant questions to nature, functionaries of the petroleum companies begin the process with scripted answers. It is not surprising that reading their own reports is very similar to a Kafkaesque novel wherein the verdict is handed down before the charges against the defendants are entered, evidence taken or the trial conducted. As Krugman states in his article, these big business-driven campaigns are "the real world equivalents of the Academy of Tobacco studies in the movie *Thank You for Smoking*, whose purpose is to fail to find evidence of harmful effects."

Despite the egregious and nearly self-evident lack of ethical and intellectual integrity demonstrated by these businesses, their schemes have been lucrative for their managers and shareholders as evidenced by their escalating profits. Moreover, hundreds of millions of dollars per year in free advertising through multiple media outlets are realized by these companies and their agendas due to the publication of these fictoids in mainstream media as well as in tabloids. Under the hollow rubric of "balanced journalism," fictoid or factoid-laden reports, as well as factual or totally fictional stories are published in newspapers, magazines, on radio and television and through the internet regardless of their meritorious or nonsensical content. In Krugman's article, a 2003 study by Maxwell Boykoff and Jules Boykoff was cited indicating that nearly equal space was given in leading newspapers to reports both espousing and discounting global warming. This was found to be the case even through the articles denying environmental degradation were directly financed by the Coalition and/or Exxon Mobil. In addition, thousands of scientific studies validating environmental deterioration, greenhouse gas increases, ozone depletion and global warming were allowed the same space as the few self-serving propaganda articles purchased by the Coalition and other special interest companies.

These flagrant examples of corporate machinations may be shocking to the conscience of many citizens on a *prima facie* level; however, similar scenarios can be recounted from most any area of business activity. Corporate predation on people, animals, places and things has a long and ignominious past, present and probable future in that a fundamental business premise is to maximize profits regardless of collateral damage. Episodes showing these outcomes throughout U.S. history have not diminished at the beginning of the 21st century. For every outrageous escapade of corporate excess and unbridled greed by the robber barons of the

19th century, commensurate examples exist of corporate *hybris* and general disdain for social and environmental damage caused during subsequent decades. The wanton excesses of the 19th century robber barons were legitimized, then enshrined as private trusts that were euphemisms for oppressive industrial monopolies. James J. Hill and Cornelius Vanderbilt assembled vast railroad monopolies, Andrew Carnegie built ironclad trusts in the steel industry and J.D. Rockefeller became one of the wealthiest men in the world by the end of the 19th century from his petroleum transportation cartels. Business practices that included stock manipulation, pay-offs to petty and powerful politicians, interlocking boards of directors and coercion resulted in a society dominated by colossal and controlling corporations.[58]

From the annals of the robber-baron reign came shocking stories from journalists that included the notorious Ida Tarbell who earned the scorn of the bloated barons and the epithet of "muckraker." Graphic accounts of sordid activities within nearly every corporate bastion were exposed in speeches and in print. Even the appalling story of Hedi Green, the richest woman in the world in the year 1900, was recounted showing nearly unthinkable avarice and callousness. It was reported that Hedi Green refused to pay a physician for medical treatment of her son's broken leg. Despite her fortune of $100 million, her son died from complications of the serious injury that could have been repaired for $15.

Although horrifying as an expression of greed and willful negligence, this mentality was displayed across a broad range of industries that affected millions of lives. Upton Sinclair's expose of the food industry in his historical novel *The Jungle* detailed the gross disregard for ordinary precautions to insure a reasonably uncontaminated food supply. Rats and other vermin along with insects and discarded offal were commonly found in food-processing vats. The corporate value network that allowed maximized profits for the food-producing corporations preempted and eclipsed all other values. Voices from individual workers or unions that objected were silenced and governmental agencies that should regulate these industries were either non-existent or impotent. The three branches of government, including the judicial, legislative and the executive offices, either acquiesced to the existing corporate value network or enabled corporate activities through enactment of rules and regulations supporting the model of *laisse faire* capitalism. This has translated often into *de jure* and *de facto* collusion by all branches of the government.

United States Supreme Court rulings upholding corporate power continued into the first decade of the 20th century and the executive branch rarely vetoed legislation and continued pandering to corporate agendas. The Supreme Court upheld the constitutionality of invidious discrimination, children working in mines and factories, production of contaminated food, unsafe medications and the wanton destruction of the environment on nearly every front. Sewage, packinghouse waste and toxic chemicals that contaminated waterways and choked the citizens were sanctioned since an effective judiciary, congress or president did not specifically prohibit them. Because of the severity and extent of environmental destruction caused through these breaches of responsibilities by the governmental branches, their continuation over many centuries discounts inadvertence or incompetence as an excuse. In fact, these patterns have been blatant to the extent that collusion between big corporations and governmental branches emerges as a plausible hypothesis in the form of a "capture theory" that has been articulated since the early 1950's. This explanation incorporates the ongoing patterns of governmental bias in favor of the wealthy corporations and concludes that the same governmental branches and agencies established to regulate corporate excesses have actually been captured by those corporations.

A case giving credence to the capture theory as a credible working hypothesis and a common corporate value network of cost-benefit analysis to explain unbridled corporate power is *Grimshaw v. Ford Motor Co.*; the Ford Pinto vehicle produced by Ford Motor Company was introduced into the marketplace in the early 1970's.[59] Litigation against Ford Motor Corporation, already having a net worth of $7.7 billion in 1976, arose from crashes that caused the Ford vehicle to burst into flames. Specifically, the Pinto model gas tanks ruptured when hit from the rear at collision speeds over 21 miles per hour; the fuel was then ignited by any spark generated from the colliding vehicles. In addition, the typical structural distortion of the Pinto vehicle upon rear-end impact deformed the door jams, thereby making it nearly impossible for passengers to escape the burning vehicle.

During the course of the litigation it was established that Ford Motor Company was fully aware of these characteristics during crash tests prior to commercialization. However, the company refused to spend $11.00 per vehicle to install a rubber bladder in the fuel tank, remove a small flange and reduce the length of several exposed bolts that punctured the tank upon impact. These very inexpensive modifications would have prevented the fuel tank failure, thereby eliminating the dangerous design flaws that had been known to Ford's engineers during the pre-production testing. In addition, J. C. Echold, automotive director of safety for Ford, opposed the National Highway Traffic Safety Administration's efforts to strengthen the already inadequate gas tank safety standards by arguing that such a proposal would be too costly. In a self-serving report written by Ford entitled "Fatalities Associated with Crash-Induced Fuel Leakage and Fires," it was argued that making a fuel tank more safe would not show a cost-benefit to society.[60]

During trial and in appellate court, Ford used a study produced by the National Highway Traffic Safety

Administration itself that indicated a maximum $200,000 per death could be imposed on any automotive corporation for producing a vehicle with a fuel tank that ruptured upon impact. The statistical analysis made on this cost-benefit schedule concluded that the maximum that Ford would have to pay would be $50 million dollars for death and injury claims. Costs associated with repairs totaling $11.00 for each vehicle would exceed the $50 million in the aggregate. Therefore, Ford refused to repair the 11,000,000 Pinto vehicles that had been produced with a potentially explosive gas tank and frame distortion (Appendix K).

Despite the continuing tragic consequences of Ford's refusal to take remedial action to prevent further major injuries and death, the National Highway Traffic Safety Administration also refused to revise its automotive manufacturing standards. As a result, its inherently dangerous vehicle along with others such as the GMC's Corvair, described by Ralph Nader as "dangerous at any speed," were sanctioned by the Federal agency established to regulate the corporations producing them. These kinds of cases, common to every industry, supports the capture theory wherein a regulatory agency acquiesces to corporate power and enables corporate predation rather than protection in the name of profit. Additionally, the tax incentives for all levels of government, wherein tax revenues increase commensurately with corporate profits, create conflicts of interest and serve to decrease governmental control of corporate excesses.

Without the safeguards of countervailing authority adequate to protect the citizens as consumers, the business model promoting profit regardless of damaging consequences has the potential to dominate the social landscape. Because there is no constitutional prohibition limiting corporate profit *per se*, the only effective control resides in the regulation of the means used to maximize profits. Legislation that has established Federal agencies to control corporate activities has also specified lawful and ethical safeguards placed on corporate monopolies and price-fixing that are in restraint of business trade and commerce. Similarly, courts of proper jurisdiction have been duly constituted and authorized to adjudicate cases and controversies arising from the application of regulations to corporate practices. However, the increasing number and magnitudes of corporate excesses, unchecked by any of the legislative, judicial or executive branches, provide mounting evidence that a business-driven profit model dissolves most opposing legal or ethical models.

When these corporate priorities translate into business practices that affect society on multiple fronts, without adequate government protection only a limited number of alternatives remain to safeguard citizens and their environment. One potentially viable alternative is founded on the proposition that a corporation has a conscience and/or ought to have a conscience. This independent but overlapping premise has been advanced as a basis upon which needed restraints are placed on corporate self-interests, pursuit of profits only and greed. It is postulated as a model of personal conscience wherein the individual person as moral agent experiences the uncertainty and anguish associated with an authentic moral decision. Rather than accepting a corporation or a person making a decision as a mere exercise in mechanically grinding out scripted conclusions, ascribing conscience to a corporation imposes standards of responsibility, accountability and obligation to that entity. Justification or rationalization for decisions and behaviors of a corporation bound by the operations of conscience loses credibility when corporate actions are shocking to a socially well-formed conscience shaped within a moral point of view.

The dynamics and elements associated with the concept of corporate conscience are varied and have been identified in various ways and from several perspectives. Generally, in the history of ideas from both the Oriental and Western traditions, law rather than ethics has been looked upon as the arbiter of collective rights and obligations. The law, also, has been the sole source of authority to enforce rules and regulations that govern corporations. Ethical standards and moral behaviors have traditionally been relegated to ecclesiastical authority and church institutions that have formulated a broad range of sanctions for breaches of their commandments. In addition to otherworldly penalties or rewards in the after-life, churches have shown creativity in devising means to force compliance with their mandates. Acceptance into a church community, sustaining members in good standing or elevation of persons to a hierarchy of church positions are bestowed on worthy members but are withheld from those deemed unworthy. Moreover, church denominations over the centuries have been proactive and creative in devising means of ostracism, shunning and punishment for violations of their parochial norms. Puritans branded adulterers with an indelible letter "A" and locked malefactors in stocks on the commons, while others were burned at the stake, drowned or "pressed" under heavy stones until dead. Within the context of these institutions, the formation of a conscience with its outer displays of behavior were defined and enforced by organizations.

However, organizations, including religious, business and government, have increasingly undertaken courses of actions that have profound ethical and moral consequences. With legal oversight and enforcement diminishing, along with legislation and court purview being eroded by corporate influence, the need for ethics operating within a collective organization becomes increasingly urgent. In order to become a viable countervailing power to corporate predation on the citizen, it needs to be shown that a collective organization such as a corporation can be a moral agent. However, one of the elements commonly thought to be essential to ethical choice and responsibility and moral agency is that of personhood.

This position is articulated by Peter Drucker, a

pioneer in 20th century business management and ethics, in the following:

> Ethics is always a matter of the person. But is this adequate for a "society of organizations" such as ours? This may be the central question for the philosopher of modern society in which access to livelihood, career and achievement exists primarily in and through organizations-and especially for the highly educated person for whom opportunities outside of organizations are very scarce indeed. In such a society, both the society and the individual increasingly depend on the performance as well as the "sincerity" of organizations.[61]

The controversy could be partially settled by indicating that a corporation has been traditionally recognized by the law as a "person." Even a "corporate sole" as legislatively enacted in some states, consists of only one person who serves as the corporate officer and the acting board of directors. While legally considered a person, those operating and owning the corporation are insulated from most forms of liability and personal responsibility by the protections afforded by the "corporate veil." A preponderance of evidence in civil law litigation or presentation of evidence beyond a reasonable doubt is needed in a criminal proceeding is needed to penetrate this corporate veil for successful prosecution of individuals within the corporation.

These distinctions and concerns, summarized by Drucker regarding the central problems of imputing moral responsibility to corporate actions, make it difficult to assign free moral agency to a corporation. If a corporation is considered a person, then it appears that it is no different from any other individual moral agent. If not a person in any meaningful sense, other than as a legal convention, it is problematic to hold a corporation to any relevant or reasonable ethical standards. When this quandary is combined with the traditional operations of the business model over many centuries, the prospect of developing a credible environmental ethic within a business environment is uncertain. It remains to be seen whether a plausible case can be made for building an environmental ethic within a society dominated by corporations, most of which produce and use multiple types of technology.

Chapter Ten

Technology, Organizations and a Credible Environmental Ethic

The complex questions surrounding the feasibility of developing a credible environmental ethic are further compounded by the proliferation of technology within societies increasingly dominated by organizations. These organizations have multiplied in numbers and types generally, and specifically have become more complex even within traditional organizations such as educational institutions, churches, governments and businesses. Within each of these categories and organizational structures, regulations, personnel and inventory of all kinds have expanded such that communication and informational infrastructures are constantly being enlarged and refined just to maintain internal cohesion. Organizations large, medium and small require incrementally more and better methods to sustain their own activities and exponentially more to expand or to interact with other organizations. A business that produces technology-based goods and services is required to become more sophisticated and productive of innovations within a shorter timeframe. The demands imposed on technology-based companies that are associated with computer goods and services have had to respond to the relentless pressures imposed by Moore's

Law. This remorseless imperative has been translated into a technology drumbeat since 1966 when it was stated that computer capacity doubles every two years. Those who have not heeded this pronouncement made over 40 years ago and shown to be prophetic have experienced an untimely demise.

Organizations that do not produce a specific inventory of technology goods and services are required to use the technology devices, services and networks to remain viable. Computer capacities that extend into a wide range of business functions have become a rising technological tide that affects vessels of all sizes, including the ship of state. Internet access has become a standard and an operational requirement for most all entities involved in business competition and its ancillary capabilities represent survival tools in a highly competitive arena. Technology and its proliferation in all sectors of society are both pushed into use by creation of needs and are pulled by the users through their demands based on perceptions of need. This ongoing cross-pollination fuels the rapidly increasing development and commercialization of technology products in all markets. Technology inventory as a commodity is developed, manufactured and marketed as any other fungible item that is to be sold in the marketplace. The dictates of this forum determine whether any commodity survives, thrives or dies; this "invisible hand" of commerce makes the decisions rather than alternative centers of authority.

Such a description of the interactions among technology, organizations, money and marketplace priorities would leave little room for competing voices in this tightly wound and welded world of economic determinism. The "iron laws of economics" that define all relationships and establish all values reducible to a cash nexus give the imperialism of technology and corporate businesses pride of place in contemporary society. Despite the recognition of a corporation within legal language as a "person," many of the common characteristics or traits of an individual person may appear absent in corporate organizations. Personhood has traditionally been associated with individual responsibility, rational behavior and personal decision-making processes within the context of ethical deliberations and moral actions.

All of these fundamental elements have been difficult to find or clearly identify in corporate cultures that encourage individual acquiescence to organizational policies, procedures and protocols. Given the primacy of corporate profit seeking, the hierarchical layers of individual workers are defined in terms of their achievement of these company goals. The quest of profits in itself is an amoral objective and the functions and means of reaching them are equally morally ambiguous and without inherent ethical content. As organizations become more specialized, functions of policy, procedure and protocol formations have increasingly been relegated to cadres of managers who establish networks that most efficiently reach company financial projections. This network itself displays a morphology common to most complex organizations that culminates in a bureaucracy with its own unique set of characteristics. This network of bureaucrats makes their organization increasingly opaque, impenetrable, lacking in personal responsibility and fundamentally amoral. When these advanced bureaucratic organizations are entwined with technology goods and services that are also inherently amoral, a credible case for corporate ethics becomes increasingly difficult to make.

In order to overcome at least some of these obstacles to imputing the functional equivalent of personal conscience to a corporate organization, the necessary elements of a moral conscience need to be analyzed. A threshold requirement to achieve this objective involves the intellectual and emotional processes of ethical deliberation undertaken by a person making a moral decision. Initiating such a process and beginning such a deliberation has been characterized as that of taking a moral point of view. Generally, every discipline demands a set of terms and conditions that are necessary and sufficient to participate in that subject matter. Language itself needs to be shaped to accommodate communication and information transfer within various subject matter areas some of which develop their own lexicons to define words and phrases. In doing so, every area of discourse strives to achieve both correspondence and coherence of its language with the expressed and claimed realities of the subject matter.

In the area of ethical discourse, an authentic moral agent needs to adopt a moral point of view wherein those similarly situated are able to process issues, share information along with values, and come to reasoned conclusions.[62] By so doing, those taking a moral point of view necessarily commit to a limited set of shared premises, including being responsible and accountable for the decisions made from this common vantage point. Others similarly situated within this perspective are able to discern and recognize that all participants are also responsible moral agents within this decision-making context. Despite the seemingly obvious meaning of responsibility or accountability, there are several distinct ways that these terms may be understood, especially when subjected to linguistic analysis including strict scrutiny of their denotation and connotations.

Three of these ways of understanding responsibility and accountability include that some person is to blame, some specific action is to be taken or a pattern of trust-worthiness is expected. The terms used by ethicists to categorize these tripartite meanings are the causal, the rule-following and the decision-making categories. The statement "a person is responsible for those actions" implies a causal nexus between the actions and a specific agent. "A physician is responsible to preserve a patient's health" is an example of the second category of meaning wherein moral responsibility and accountability requires adherence to established rules such as are found in the Hippocratic Oath. "That person is (will be) a responsible person," imputes an expectation of a morally acceptable capacity, a willingness to make decisions and to be accountable for them.

These distinctions may serve to clarify the ostensible meanings of these terms in professional and ordinary language discourse, but it remains problematic to apply these meanings to a corporation itself or to persons acting within the context of a corporate organization. Within a corporate context, the organizational structure, including organization flow-charts, chain-of-command guidelines along with rules, regulations, protocols and procedures, creates a "corporate culture" that fosters the first two categories of responsibility/accountability. Employees within a corporation are expected to follow rules, often slavishly, and causal connections regarding the operations of cause and effect typically operate in most all corporate organizations especially when employees are carefully monitored and supervised. Typically, systems of rewards and punishments known to everyone within the organization reinforce compliance to responsibility and accountability on the two levels of meaning that include causation and rule-governed behaviors. For these two categories to operate effectively within the context of a corporation, the structure of the organization needs to allow a degree of transparency sufficient for compliance evaluation.

Even though these elements may be in place

within a corporation, the additional meaning of responsibility/accountability understood as a decision-making process is required of an organization to participate in a moral point of view. A multitude of voices over the centuries have provided further clarification and articulation of this decision-making process; however, a useful summary of several influential traditions is provided by William Frankena in his book *Thinking About Morality* written in 1980:

> A morality is a normative system in which judgements are made, more or less consciously [out of a] consideration of the effects of actions…on the lives of persons…including the lives of others besides the person acting…. David Hume [1711-1776] took a similar position when he argued that what speaks in a moral judgement is a kind of sympathy…. A little later Kant [1724-1804] put the matter somewhat better by characterizing morality as the business of respecting persons as ends and not means or things.[63]

In this summation, the venerable tradition of rationalism in its multiplicity of iterations is emphasized as fundamental to a moral point of view, as is the basic emotion of benevolence with respect for persons and duty. The tradition of moral deontology is derived from the Greek word meaning duty that defines the imperatives of Immanuel Kant's ethical position. His creative combination of rationalism and deontology resulted in the formulation of the categorical imperatives that give guidance and context to moral behavior. Kant's formulations of the categorical imperatives provide both form and function for those taking a moral point of view by focusing on common denominators of the human condition, rather than only on individual preferences. That all moral agents act such that the principle of the action can become a universal law without self-contradiction expands the moral deliberation from a merely egocentric standpoint to one encompassing all of humanity. Lying, for example, would be a violation of this formulation since a liar is parasitic on the truth-teller and would contradict the central functions of communication if universalized as a principle of action.

The basis of these expanded ethical parameters is recognition that other persons have intrinsic value and their lives are of inherent worth. To think or act otherwise would be self-contradictory for those living in communities as social beings. Because of this commonality of value the "kingdom of ends" formulation of the categorical imperative can be logically inferred and made central to a duty-driven life. Within this context, discharge of a moral duty is done "come what will" without weighing or measuring consequences. The third formulation of the categorical imperative as a guide for moral decision-making involves transforming the principle of one's moral action into a uniform law of nature without self-contradiction. One example would be the wanton destruction of the environment or, by implication, any arbitrary destruction of life forms that would create a contradiction if made into a uniform law of nature. If such a principle were uniformly applied as a law of nature, life itself would cease, thereby resulting in a self-contradiction.

Despite the appeal of giving structure and content to the moral point of view with the specific formulations of the categorical imperatives, their applications to corporate organizations, then to an environmental ethic require additional considerations. Although there may be a multiplicity of obstacles in doing so, the plausibility of this position suggests that many substantive objections can be met. One common objection to imputing conscience to a corporation is that a corporation, in seeking to maximize profits, will make exceptions to ethical imperatives for itself. While a corporation may pay lip service to minimum moral business standards, it will not abide by them and will, in fact breach them at will. However, Immanuel Kant, himself a rigorous deontologist, recognized that individuals themselves are also prone to set aside the obligations of ethical imperatives in an effort to gain personal advantage. Based on these premises showing little difference between personal and corporate arbitrary suspension of the ethical, both share in the common denominator of hypocrisy. Hypocritical behavior is not the exclusive hallmark of corporations, but disjunctions between professed standards and intended or actual performance are manifested by almost all persons and their social groups, including church organizations.

Whereas the corporation as a complex

organization may be considered lacking an obviously discernable conscience as a moral compass, it has certain central attributes in common with persons. Both persons and the corporate organization are embedded within systems that create their own sets of imperatives. A person, for example, is born into a multiplicity of systems that include family, caste, class, governmental, educational and religious.[64] Each of these systems immediately and systematically imposes rites, rules and regulations that demand obedience and compliance to the requirements of those systems. Similarly, the corporate organization is enmeshed within many systems, some of which it inherits and some of which it helps to construct or reinforce. In this systemic context, both the individual person and the corporate organization need to negotiate with systems which often compete for attention and authority. Because of these parameters, it is not sufficient to simply demand that corporations reflect the ethical principles or moral behaviors of an individual that are projected on to it "writ large." Projection of any one or all of the meanings of individual responsibility onto a corporation does not produce unequivocal attributes that can give ethical guidance from a moral point of view.

Rather than naively projecting the demands of the categorical imperatives or any other logical extensions onto a corporation and holding it to those standards, the direct implications and nuances of the relevant systems need to be considered. Traditionally, the primary and most influential systems into which the corporations are emersed have been stated in terms of "two hands." On the one hand, there is the "invisible hand" referenced by Adam Smith in his *Wealth of Nations* published in 1776 and the other is the hand of government. According to Smith and his proponents, the "invisible hand" represents the workings of multiple hands toiling in a free and competitive marketplace without governmental interference or moralistic meddling. Free of these projections, manipulations, distractions or coercions, the spectre of self-interest rules a kingdom wherein the content of conscience, along with moral responsibility and accountability become the outcomes of the competitive struggle in an arena of beings calculating their own advantage. In his other writings regarding ethical topics, however, Adam Smith shows his fidelity to the 18th century Scottish School of "benevolence" that included the influential David Hume. As a moralist, Adam Smith urged benevolence and "sympathetic feelings" for one's fellow pilgrims on the journey but concluded that only unfettered competition and unbridled self-interest were enough to control both individual and corporate behavior.

The "Smithians" claim that ethical projections or preaching, including moralistic Jeramiads or reverent rantings, were ethically impotent, ineffective and irrelevant. Within this system, abstract or applied concepts of justice or fairness would be replaced by corporate strategies and economic advantages as relevant topics in a corporate boardroom. Similarly, discussions that would parse out categorical imperatives, natural law or doing the right thing behind the "veil of ignorance" would be exercises in folly and futility for the decision-makers in the corporate world. In fact, the contemporary proponent of using the "veil of ignorance" behind which just and fair moral decisions can be made must recognize the difficulty of applying this construct unequivocally to corporate behavior.[65]

This "veil of ignorance" construct was proposed by John Rawls as a means of shrouding self-interest and strictly self-serving considerations from the decision-makers to allow considerations of justice, fairness and benevolence into the discussion. Devising even hypothetical models for individual moral choices is difficult and constructing an actual system utilizing the "veil of ignorance" requires extreme resourcefulness. When applied to corporate ethics, these difficulties are compounded primarily due to the complexity of the multiple systems involved within the corporation, its agents and the many congruent and incongruent systems found in societies worldwide. It is not clear, for example, that the mixed use of ATV's or snowmachines in wilderness areas or clear-cutting of an old growth forest would be different on either side of the "veil." (Appendix L)

Because of the complexities arising from the increasingly complex systems, Rawls himself recommends focusing on "reflective equilibrium" wherein individuals and corporate organizations experiment with hypothetical "micro-situations" to determine decision-making methods. Those experiments that are deemed most successful will become the working models offering guidance for actual moral decisions. While creative and innovative, Rawls' suggestions are fraught with perils and uncertainties even when favored models are used in real life and for real-time moral decisions.

Although the "hand of government" has often been explained within the context of a social contract model, the application of ethical terms, such as justice and fairness to adjust or limit government actions, raises additional objections. Both Rawls' characterization of the roles of justice and the hand of government position are based on a mistrust of human nature expressing itself in the marketplace. Both look to the binding forces of contracts among parties to bring about individual and corporate well-being within an equitable commonwealth. The suspicion harbored by the "hand of government" advocates toward the "invisible hand" proponents has been articulated by John Kenneth Gailbraith, a leading social and political economist since the early 1960's. According to this popular position, strong governmental programs are necessary to protect the interests of citizens against the rule of the invisible hand controlled by the wealthy and powerful corporations.[66]

Despite the palpable differences between the invisible and visible hand social theorists, both recoil from the imposition of ethical theories or moralistic teachings onto corporations. Both repudiate the intrusion of an ethical lexicon or applied moral language into the fabric of corporations and their agents from sources external to the operations of the corporate systems. The systems into which the corporations are emersed, enmeshed and embedded manage, adjudicate and define appropriate ways of behaving. Inappropriate actions, on this view, are not defined by adventitious or external moral sources but are matters of internal management within those systems.

For the leading advocates of both the visible and invisible hand positions, the necessary and sufficient moral guidance for correct corporate behavior resides within the incentives available to them by these systems. For those urging increased activity by the visible hand of government, the political and social programs imposed by government, along with their legal and economic controls, become the mechanisms of morality, not the preaching from tuned or untuned pulpits. Similarly, the invisible hand of the unregulated activities of self-interest is thought to preempt moralizing from outside oracles and eclipse ethical theories or moral pronouncements through the construction of a system of rewards and punishments, and pleasure and pain generated from the operations of marketplace activities themselves.

As a result of their positions, every ethical theory or moral action, including the environmental ethic, becomes a function of a system dominated by self-interest, amoral technology, the workings of economic self-interest and the legal processes promulgated by bureaucratic administrators. On these foundations corporate organizations embedded in complex systems and their minions have transformed traditional ethical teachings. Such transformations have moved the basis of ethics and moral actions into radically different sources of ethics from which moral authority is derivative. These paradigms have shifted standards of ethics into complex systems of political correctness, have made categorical imperatives into obedience to arbitrary and self-serving regulations, and have replaced inalienable ethical values and civil rights with cost-benefit calculations and have changed corporate agents with an authentic conscience into hand puppets manipulated by both visible and invisible hands.

CONCLUSION

Conclusions regarding the interactions of technology and environmental ethics need to remain tentative. They also have to reflect the multiple factors that contribute to the increasing complexities developing within the dynamics of societies and cultures. Some of the factors are based on facts that can be established with scientific experiments and empirical verification; others are factoids or fictoids that may contain elements of fact but lack sufficient factual content upon which firm foundations for thought and action can be built. Such is the predicament foisted on citizens when centers of authority, whether lawfully held or seized by power, "tune the pulpit" with fabricated and/or manipulated factoids.

Ethical propositions or recommended moral actions promulgated from these sources of propaganda are porous and mirror self-serving, distorted and ultimately inauthentic counsel. Corporations, religious sects, political parties and a multiplicity of special-interest groups have shown resourcefulness in generating cascades of bogus pronouncements that serve only their agendas. Real facts, foundational to ethics, moral action and their consequences, are eroded, distorted and are shamelessly denied in the narrow focus of the parochial and powerful seeking their exclusive self-serving interests. As such, this orientation transforms ethical discussions into dogmatic assertions without factual content, manipulates moral behavior and reduces dialogue to strident diatribes.

Within such a fractured and fragmented landscape, the elements of reason, the common good, social and distributive justice, fundamental fairness and the moral point of view are eclipsed. The threshold requirements of an authentic rational ethical system such as proposed by Immanuel Kant cannot be met against a cacophony of self-serving voices. Even Kant's general epistemological observation relevant to an ethical context that "concepts without percepts are empty and percepts without concepts are blind" is lost amid the self-serving propaganda emanating from a contrived Tower of Babel. Without a reliable and verifiable set of relevant facts commonly accepted, the percepts are unable to give factual content to our collective ethical concepts. Similarly, without the common grounds of reasonable concepts that give shape to ethical discourse and moral action, the percepts, as collections of some facts, factoids and fictoids are blind, thereby providing little ethical guidance and a poor moral compass.

Those who search for ways to effectively project the elements of individual moral conscience or virtue onto a corporation are challenged to overcome significant obstacles, especially when the stakes are high in all areas of ethics and moral behavior, including the environmental ethic. Informed and conscientious parties to the discussions recognize that most all societies are dominated by corporations by almost any measure;

however, strong historical cases can be made that neither the visible nor the invisible hands have been able to make corporations meet even minimal ethical standards. Campaigns initiated by both "hands" that use advertisement in a broad range of media outlets to manipulate behavior under the auspices of morality have produced slogans but little else. None of these slogans have given solid evidence of a corporate conscience at work.

Even environmentally friendly sounding slogans, devised by corporations such as General Motors or Ford Motor Company promoting non-petroleum fuels, mask the reality of their gas-guzzling automobiles and their consequences. The production of small numbers of "flex fuel" vehicles capable of using petroleum and/or ethanol-blended fuels was for ulterior motives that were sanctioned by the EPA. Production of a token number of flex fuel vehicles enabled Ford and General Motors to inflate the average miles per gallon APMR ratings for all of their vehicles with impunity and with the collusion and legal sanctions of the EPA.[67] In these cases that have resulted in continuing the depletion of non-renewable resources and increasing pollution, neither the government nor the marketplace has provided necessary environmental safeguards. Under these chronic conditions of marketplace and governmental failures, the need to bring corporations into serious ethical discussions and into a moral point of view arena is increasingly urgent. With a corporate conscience as a moral compass established by projection that is based on duty, consequences or by contract, and the commensurate demands on individual conscience, dialogue regarding a meaningful environmental ethic may be possible.

However, any concerted effort, no matter how organized or implemented to integrate elements of a moral point of view into corporate and government decisions, confronts multiple obstacles. For those urging the establishment of a viable environmental ethic based on the imposition of conscience on corporations and governmental agencies, there are daunting challenges involved with even proposing such a goal. The linkage of fundamentally amoral and non-ethical entities such as sovereignty, nation-state, political administrations, business models and corporations forms a formidable fortress that militates against an environmental ethic. When combined with amoral science and technology programs legitimized by legislators and courts of law, business and governmental alliances appear to dwarf and dissolve all opposition to their agendas. The concepts of sovereignty, science/technology, business interests, political agendas and the rule of special interests rather than the rule of law have eclipsed traditional concepts of the environmental ethic such as justice, fairness, common good, sustainability and responsible stewardship. This is especially apparent when these concepts are filled and mobilized by the percepts of historical events.

These difficulties are compounded when the central

issues are involved with decisions that derive ethical guidance from collective rules and when moral decisions are based on those derivations. Any illusion of certainty entertained regarding the simplicity of making moral choices based on collective rules is quickly shattered when subjected to strict scrutiny. Conscience formation based on general *a priori* premises or on specific *a posteriori* experiences remains adumbrated even when clarity of conscience is sought through rigorous moral discipline. Even residents of a cloistered monastery recognize the challenges of developing a consistently well-formed conscience and acting pursuant to it within their measured quotidian lives governed by rites, regulations and general rules.

Despite the clearly defined basis of authority and rules in such a community, it is still problematic to make authentic moral decisions in this context without lapsing into a slavish and mechanical way of life. However, Sigmund Freud's analysis of conscience as sublimated and internalized authority figures begs the essential questions relevant to devising an individual ethic and making authentic moral decisions on a daily basis. The ethical dilemmas and moral predicaments are not abated by a recommendation that one's internalized authority figures must rule. Any process or criterion for selection of authority figures to internalize and obey involves the construction of indices and matrices to establish these processes and criteria. Since it is not intuitively clear as to the most favored or dominant authority figures that would best form a conscience, both individuals and collective organizations would still be left without a reliable or rational basis for conscience.

If conscience is to be useful to either individuals or collective organizations for moral decision-making, a rational and reliable basis needs to be established. In the area of environmental ethics, multiple parties to environmental controversies have frequently exploited the malleable aspects of conscience. The strong "proclivity for rationalism and idealism," mentioned by Karl Rahner has favored deduction of ethical conclusions from *a priori* premises. Usually an individual or collective conscience shaped on a rational and/or idealistic platform would seek authoritative *a priori* premises as a foundation.

A result of this orientation toward deduction on the part of the wilderness conceptualists is that their conclusions remain idealistic and conceptual. Each issue, no matter how concrete, particular and practical, is decided by derivation from those *a priori* premises. As a logic-driven exercise, the wilderness concept arguments terminate in conclusions that are already explicit in the major and minor premises of the logical syllogism. This is not unique to the wilderness conceptualist arguments within environmental ethics, but is the outcome of any position logically deduced from *a priori* premises. When entirely devoid of empirical or factual "percepts," conclusions from these premises are the logical equivalent of mathematical tautologies. As definitional and as functions of rigorous logic, these conclusions are unassailable as a matter of internal coherence. However, this methodology fails to establish correspondence with the empirical world of percepts or facts that indicate that the wilderness concept does not exist in that world or its universe of discourse. (Appendix M)

The further insistence of those espousing and defending the wilderness concept in controversies related to environmental ethics on entirely uncompromising premises, makes the separation from public policy consensus-builders even more pronounced. This expanding gulf among the disputants makes the abyss of disagreement more deep and wide. It also makes the prospect of building a bridge to connect the dialogue and bring about a modicum of détente more unlikely. For those engaged in practical public policy formation and its implementation, the unqualified adherence to immutable wilderness concepts is considered little more than a way of hiding behind dogma, nostalgia and romanticism As a wall, these elements have two sides, one that walls them in and the other side to wall all others out. With this model in place and with those on each side of the wall seeing the opposition as inhibiting meaningful dialogue and as inhabiting the refuge of scoundrels, cracks in the walls may paradoxically prove to be one means to begin the crumbling and erosion process.

One such significant crevice in the wall of environmental ethical separation is found in common human behavior patterns that have directly impacted all idealistic, romantic, nostalgic, rational and deductive systems of thought. It is specifically that large numbers of those who profess allegiance to these *aretean* standards do not, in fact, live up to them. A multitude of historical examples that illustrate seemingly impenetrable fortresses of idealism, romanticism and hardened dogmatism also show that the practice of them is often honored in the breach. The patterns of human behavior that evince this discrepancy have been characterized in various ways, including human weakness, ignorance, hypocrisy, fraud, denial, delusion and deception, including self-deception.

These ideologues who base their position on ideal concepts and profess an uncompromising adherence to those ideals espouse *aretean* standards that are indefeasible, unqualified and unconditional. Often that position is further reinforced by inappropriately comparing their utopian world to the real world of the opposition. Such a strategy dramatizes the distinctions and may result in a debater's victory but it follows the philosophical bait and switch tactic of the classical Sophists who made the weaker argument seem the stronger. In most any context, the airbrushed ideal will appear more desirable than unvarnished reality scuffed and tarnished by the wear and tear of gravity. Such contrived maneuvers essentially create an even greater disparity of disagreement among those engaged in discussions regarding environmental ethics. However, the exegencies of policy formation and mixed-use planning obtrude themselves at several levels including the mundane but powerful level of money. On this plane, professed idealists embodied in institutions and their individual investors may urge investments in socially and environmentally responsible companies. A list of these investment funds proposing to be socially and environmentally conscious and responsible include Ariel, Pax World Balanced, New Covenant Growth, Calvert Social Investment Equity, Parnassus Equity Income, Neuberger Bernan Social Responsibility and Vangard TRSE Social Index that, combined, have assets over $10 billion.

These investment vehicles theoretically allow the idealists to "walk their talk" by investing to the level of their convictions and financially benefiting only from environmentally responsible enterprises. As a stockholder, the investor may also try to exert its influence on corporate decisions that could be considerable, especially when wielded by large institutional investors. Such a combination of self-interest and financial pressure on corporate board decisions would theoretically bring about mutual benefit including social and environmental improvement.

Despite the apparent reasonableness of this plan "on paper" along with its potential to promote the common good, actual investment patterns often suggest a departure from such idealistic common grounds and common goals. A representative example of this disconnect between stated ideals and actual practice can be found in the investment by the Christian Brothers Investment Services which manages assets for Catholic Church institutions. Although professing social responsibility, just stewardship and commitment to the environment, the CBIS had purchased 20,000 additional shares of Exxon Mobil in 2006. In addition to the obvious dissonance between the client Church teachings regarding social distributive justice and responsible environmental stewardship, John Wilson, the Director of the CBIS has also acknowledged Exxon Mobil's significant contribution to environmental pollution, greenhouse gases and global warming. The overriding motivation for the additional investment in Exxon Mobil was maximization of CBIS's return on investment. The Director states "We invest in Exxon Mobil because our managers think that it is going to make us money."[68]

Despite the disclaimer that CBIS intends to take part in the governance of the company, there has been no measurable change in the course of doing business as usual by Exxon Mobil. Similarly, other national groups representing themselves to be ecologically, socially and environmentally responsible, invest organization funds in companies that continue to pollute and adversely affect the global environment. Several of these socially responsible and environmentally protective investment groups include the Sierra Club Stock Fund and the New Covenant Family of Funds both of which are the investment vehicles for internationally prominent organizations. The Sierra Club has long been synonymous with the highest and best forms of wilderness preservation agendas in the world. In common parlance, the phrase "tree hugger" refers to members who claim to be the paragons of ecological stewardship. However, McGraw-Hill, a giant publishing corporation and a ravenous user of the very trees hugged, counts the Sierra Club as a stockholder. The Sierra Club has also invested in Apple Computer that has yet to implement a viable program to recycle their computers in an environmentally responsible manner.

In the case of NCFF directed by the Presbyterian Church that officially objects to the proliferation of weapons and their environmentally destructive results, it has large investments in United Technologies (UTX). UTX is a major manufacturer of military weapons including the Black Hawk helicopters for the U.S. military. For NCFF, along with the others representing themselves to be socially and environmentally responsible, achievement of the highest and best return on investment appears to supercede the highest and best commitment to the environmental ethic.

Despite the representations of this and other socially and environmentally conscious investment funds, actual applications of their "investment screens" show them unwilling or unable to eliminate conflicts with their own stated goals. Even though a screened fund may be weighted toward healthcare or technology stocks rather than industrials, the overlap among these companies still allows large investments in environmentally and socially destructive products. As a result, the intractable network of technology and commerce militates against socially and environmentally partitioned investments even in the short-run. In the long-run, these investment funds are guided by the same criteria of all investment funds that seek to maximize return on investment, such that the profits will be taken by the conscientious and the callused alike.

When idealistic and absolute values, such as those embedded in investment fund screens, are subjected to strict scrutiny, they are not implemented intact. Moreover, this pattern has been recurrent and a common denominator in all areas of applied ethics including environmental ethics. The appropriation of even pristine ideals such as good, justice, fairness, inalienable or civil rights, natural law or social stewardship by those who make and implement public policy usually terminates in nuanced, modified or distorted applications. Even when couched in terms of absolute ethical, moral or human rights, their applications in public policy legislation have undergone significant modifications. The Air Quality Act of 1967 and the Clean Air Amendments of the 1970s, for example, incorporated idealistic values translated into the language of civil and human rights. As stated, the language of this legislation conveyed that the basis of inalienable rights was not contingent on cost-benefit analysis. The human rights reported to be protected under these Acts were the fundamental and inalienable rights of human beings to a clean and healthy environment without condition or qualification.

Although this legislation appeared to reflect undiluted and unqualified ideals stated in legal language as "rights," the actual outcomes of such legislation were to be markedly different. Soon after the Acts of 1967 and 1970 were enacted, corporations objected strenuously and effectively to the implementation of the legislation based on their costs for compliance and alleged loss of jobs. Rather than imposing the absolute bans set forth in the Acts, bills in Congress were introduced to only slightly abate rather than ban pollution of the environment. While the corporations claimed that enforcement of the Acts by implementing absolute bans on threshold levels of pollution would be financially ruinous to the corporations, the actual costs incurred by the corporations are subject to continuing dispute. Although the cost-benefit basis for corporate objections to these bans was not substantiated, the influence of corporations on governmental enforcement agencies effectively curtailed implementation of the Acts.[69]

In a retrospective study conducted by the U. S. Environmental Protection Agency from 1971 to 1981, the negligible job losses from any environmental protection legislation were more than offset by new job creation. Most of these new jobs were in cleaner industries and some were created in companies that began to produce pollution control equipment. Despite the primarily contrived and largely exaggerated costs to be imposed on business corporations to ban pollution pursuant to these Acts, pollution bans became mere pollution modification regulations that included greatly extended deadlines for compliance. The results included illusory deadlines, non-enforcement of these flaccid regulations and actual increases in the most dangerous pollutants. Increases of the most noxious and unhealthy pollutants were allowed since they were the most costly for corporations to curtail.[70] Finally, Executive Order 12201 signed by President Ronald Reagan in 1981 required all environmental legislation and regulations to be subjected to "cost-benefit analysis" before implementation. As a direct consequence, the ensuing analysis allowed the creation of pollution units wherein the most lethal pollutants were allowed to be increased when less toxic and costly pollution quotas were slightly reduced. Within the parameters of this Executive Order, intercorporation negotiations were sanctioned wherein corporations could trade pollution units in toxic environmental transactions. [71](Appendix N)

Because of these factors and historical events relevant to environmental ethics and their applications in public policy, categorical, unconditional and absolute premises do not appear feasible. When untenable as an ethical compass used in the ship of state, assertions of absolute norms appear to be honored only by the passionate few and are often honored only in the breach thereof. Carefully crafted arguments contemporaneous to the Clean Air Acts and others based on inalienable human rights were largely ignored when their propositions confronted economics, politics and public policy. Lofty ideals, no matter how well-articulated, were dissolved by the dominant corporate business solvents of money, power and consumer wants.[72] Corporations that maximize profits and political power by creating, then pandering to rampant consumerism, have shown their massive power to preempt, eclipse and quash rational ethical alternatives to managing eco-systems.

These points of conflict without apparent resolution in theory, practice or public policy application are also seen within the context of swelling global populations. The issues, methods, norms and values that separate the conceptualists and the public policymakers also create disjunctions among planned population control advocates, artificial contraceptive proponents, prolife groups and those who randomly maximize reproductive potentials. Added to the voices are those who advance pre-implantation genetic diagnosis (PGD) and assisted reproductive technologies (ART) as technologies and processes to be used in reproduction. In many discussions, disputes within these arenas become cacophonies that often only drown out the opponents' arguments by increasing the volume of speech but fail to reach rational conclusions.

At a fundamental level, certain issues such as quantity and quality of progeny need to be discussed on a global stage and solutions need to be set in place through international organizations and governments. Failure to take decisive action and succeed on a global scale will produce consequences from over-population that will soon become uncontrollable. Those urging the generalized engineering of technological outcomes of reproduction by using PGD and ART procedures are still faced with justifying their moral choices when these methods are employed. The selection of desired or undesirable traits for a child is also subject to a broad range of critiques from scientific, medical, religious, political and legal interest groups. A strong case can be made showing that the advertising and marketing media exert a powerful influence on individual choices. The media itself has the potential of ultimately setting trait selection criteria in place. Those who may not possess politically, cosmetically or ethnically correct characteristics could easily be eliminated prior to birth or conception. The objections to a rights-based selection paradigm, such as that proposed by John Locke, include substituting the stewardship of parent-child-family-community relationships as a basis for procreative choices. Although these issues continue unresolved in ethical theory, the actual choices being made worldwide are those that will decide the overall makeup of surging global demographics. (Appendix O)

While these crucial quarrels continue unabated and unresolved, the weight of numbers, regardless of selected traits, continue to weigh heavily and unevenly on global resources. The common-denominator demands to meet the basic human needs of food, clothing, water, shelter and healthcare continue to out-strip supplies. Regardless of trait selection decisions, unqualified reproductive choices to produce any kind of child have profound consequences. For nearly twenty-five percent (25%) of children produced, their basic human needs will not be met in a world of six billion humans and ten times more animals competing for shrinking natural resources, commodities and services. The fate of these children, regardless of trait selection criteria, will be affected. Fifty percent (50%) of pregnancies are unplanned and twenty-five (25%) are unplanned and unwanted, resulting in reproduction outcomes that are controlled by random, irresponsible behavior without the constraints of any rational decision-making or cogent morality. The laws of large numbers as applied to human populations, have eroded essential human qualities to provide for basic human needs and have substituted proliferation of quantity in the absence of sexual restraint.[73]

When the fundamental issues involved with quality and quantities of reproduction are placed in a global theatre, the profundity of their impact on the environmental ethic is experienced. The existential *exigence ontologic* or the demands of being come into sharp focus in the form of six billion humans inhabiting the earth at the beginning of the 21st century with 2.7 billion living in only two countries, China and India. The land mass of these two countries is approximately the size of the United States and Canada which have a combined population of 2.2 billion less than China and India. Despite penalizing couples producing more than one child, China's population continues to grow with its residual environmental degradation. Those in China complying with population-limiting regulations seek to maximize the desired traits of their only child by using either traditional animal breeding techniques or sophisticated PGD or ART biotechnology methods available to only the elite (Appendix L).

China's population continues to increase primarily due to non-compliant breeding pairs who eschew population-reduction laws, thereby adding to the weight of numbers from the enormous existing quantity of people capable of reproduction. In India, by contrast, the traditions of fertility continue largely unabated wherein the fecundity of a couple is still extolled as a moral imperative by various castes. Combined with moral teachings of some ancient and mainstream religious institutions against "artificial birth control," the historical social dynamics of this and other populations similarly situated drive greatly accelerated population growth into the 21st century.[74] Population growth predicated on these premises and multiplied on the base of six billion humans and ten times that many animals includes the statistical inevitability of teratogenic progeny. For human reproduction, three percent (3%) of offspring will have significant physical and/or mental defects with thirty-three percent (33%) of these newborns having life-threatening conditions. The infants with teratogenic defects that survive infancy usually manifest additional health problems in childhood such that fifty percent (50%) of the children in hospitals have prenatal malformations. The depletion of natural resources and the impact on the environment by reproduction is affected by both pre-natal and post-natal medical technology.

Although prenatal ultrasound, for example, may alert parents and/or physicians to a teratogenic event, the moral decision to terminate such a pregnancy is independent of the probability of infant deformity or survival. In this context, voices are urging that values and means other than those of medical technologies be considered in reproductive moral decisions. These voices specifically speak out against the priorities embedded in a society dominated by consumerism wherein fungible items in the marketplace are all commodities to be bought and sold at a negotiated price. A call to procreative stewardship resonates within a counter-cultural context for those disposed to consider other options of procreation rather than only quality controlled, stipulated or randomized reproduction. As a major factor within the context of reproductive or procreative choices that directly affect population impacts on the environment, consumerism has emerged in the middle of the 20th century to control the values, motives and the actions of large segments of the global population.

Consumerism, as a description for a cluster of related behaviors, is identified by its common denominators of both the acquisition and consumption of desired goods and services. As a prioritized set of values, consumerism focuses on the goals of purchasing and using wanted consumables within a competitive arena of scarce resources. While goods and services have been staples for acquisition, use and exchange for millennia in geo-political economics, they had been largely restricted to meeting basic needs for food, water, shelter, clothing and medicine. With the rare exception of royalty and aristocrats who seized and sustained their status by military force, the vast majority of people sought only to meet their basic needs. Opulent life styles manifested by only a few in every society were viewed as unattainable by the masses rather than held up as an achievable goal. Lack of access to resources by the vast majority of any demographic group to meet more than basic subsistence needs precluded opportunities to attain affluence.

These barriers that prevented almost all subjects of every tribe or sovereignty from purchasing luxuries were a combination of meager technology and political policy priorities, including economic mercantilism and military preference. The coercive powers of sovereigns exerting their authority through political machinations, money manipulations and naked military might blocked the development of consumerism until the mid-20th century. Led by the political, technological and economic machinery of multi-national corporations, the essential consumerism elements combined immediately after World War I, especially in the United States. Despite sporadic consumer-oriented efforts initiated by large corporations in the late 19th century, the critical ingredients of mass production, targeted-saturation marketing and discretionary income of consumers were not present in combination to ignite a conflagration of consumerism. Its initial large-scale appearance in the United States during the 1920's was suppressed but not extinguished by the Great Depression of the 1930's. In fact the widespread deprivations of those living during the Great Depression were reinforced by the limitations of rationing during World War II. These factors helped to create a political, economic and social climate favorable to corporate growth for the mass manufacture, marketing and distribution of a wide range of goods and services. With a more centralized government exerting increasing control over finances in every sector of the economy, personal income escalated to allow many discretionary purchases. In this way, the prophetic premise of Karl Marx, that the workers could not afford commodities produced by the capitalists, was radically over-turned although his cash nexus, defining the basis of all human values and relationships, was left intact and was essential to flourishing consumerism.

At this juncture, a cascade of goods and services of all descriptions flowed into the marketplace in unprecedented quantities. With all the necessary and sufficient conditions in place, the tipping point for consumerism proliferation was reached and its dynamics have produced a seemingly inexorable profusion of consumption frenzy that has increased almost geometrically rather than arithmetically. The environmental results are the direct consequences of a growing consumer *credo*, which in a mere fifty years has become an international mantra and a new categorical imperative. This belief and value system, made respectable by almost all arbiters of behavior, has fueled the irreversible and continuous destruction of our natural resources.

In this media and merchant-driven world, sanctioned by politicians, educators, churches, tribes and families, the ethical and aesthetic "good" has been "morphed" into "goods" as a term reducible to things accumulated. Any sense of "evil" within a contrast theory of meaning standing in opposition to "good" has been, in Fredrich Neitzsche's terms, "transvaluated" and/or it has become a "banalized evil" as described by Hannah Arendt in her essay *The Banality of Evil.*[75] Efforts to construct complex wholes bonded to "good" as suggested by G. E. Moore usually fail to resonate with leaders and followers alike unless defined in the operational terms of consumption. In contrast to G. E. Moore's own hierarchy, placing personal affections and aesthetic enjoyment at the summit of "complex goods," these values are replaced by the consumer litany that sacrifices those and other alternatives on the cash register of resource predation by purchases. Against these formidable countervailing forces, the reality of an authentic environmental ethic clearly depends on the values, virtues, choices and actions taken by all parties to the environmental ethics debate on an international stage by local and parochial groups to abate environmental destruction in societies overrun by rampant consumerism.

Endnotes

[1] A discussion of this topic is found in "A Theory of Environmental Virtue" by Ronald Sandler, *Environmental Ethics*, Vol. 28, No. 3, 2006.

[2] The author attended a series of lectures presented by Brand Blanshard, Professor of Ethics at Yale University, wherein he systematically evaluated multiple traditional ethical systems applied to a hypothetical case of stealing a $1,000,000 jewel from Tiffanies in New York. He showed that an individual moral agent could "easily justify" stealing the merchandise under the premises of any of the leading ethical systems. In fact, if significant need of the moral agent were present, jewel theft in this case could easily be shown to be a moral obligation since the loss would be spread over millions of premium payers of the insurance company that would actually cover the loss.

[3] St. Augustine of Hippo (354-430 BCE) recognized the distinct and profound differences between conceptual ideals and the real world of partial solutions, amelioration and compromises. In Augustine's *De Civitate Dei* (The City of God), every person on the journey is a citizen of both worlds, the ideal City of God and the real City of Man wherein the dual citizenship of everyone gives rise to irreconcilable dissonance and conflict; these problems are not amenable to permanent solution. A millennium prior to Augustine, the prophet in Isaiah 35 describes the same predicament for those seeking the Kingdom of Shalom which requires *Hanukkah,* a Hebrew word for commit-

ment or determination. In Dale Wasserman's *Man of LaMancha*, Don Quixote was driven in quest of this "impossible dream" to make the real world align with his vision of the ideal.

In addition to the prescribed moral virtues of faith (determination), hope (aspiration) and love (charity), Augustine encourages cultivation of the "brother virtues of anger and courage." These two are thought necessary to implement the traditional three cardinal virtues to cope with life in the City of Man. In the other City, only love will be of value. Of course, entertaining this "impossible dream" in any of its iterations requires commitment to appropriate virtues in our world of heavy and hard reality.

[4] These principles are included by Childress and Beauchamp in their works on bioethics.

[5] John Rodman, "The Liberation of Nature," *Inquiry* 20, 1977, and Garrett Hardin, " The Tragedy of the Commons" in *Toward a Steady State Economy,* Herman Daly, ed., San Francisco: W.H. Freeman, 1973.

[6] Garrett Hardin, "Living on a Lifeboat" cited in *Environmental Ethics,* Vol. 26, No. 4, 2004, p. 342.

[7] However, Smith admitted that capitalism's "wasteful materialism" was a significant side effect of this expression of self-interest and that the "invisible hand" allowed "...people to ruin themselves by laying out money on trinkets of frivolous utility." Smith's analysis of self-interest driving capitalism gave rise to widespread social ills noted by Thomas Carlyle who characterized Adam

Smith's form of economics as the "dismal science." In the same century, Thomas Gray's Elegy Written in the Country Churchyard (1753) poetically recounted the social results of this "dismal science" from which "Chill penury oppressed the noble rage and froze the genial currents of the soul."

[8] For example, recent litigation initiated by endigenous Arctic Peoples against corporations causing environmental pollution adversely affecting their natural food resources has been brought under the legal theory of civil rights violations. The constitutional basis of redressing gross inequities may be defensible but payments as a remedy to the plaintiffs would only spread the ill-gotten gain. Similar to tobacco cases, payments to the plaintiff only serve as the cost of doing business rather than imposing an effective ban on polluting business practices.

[9] Susan George, *How the Other Half Dies: The Real Reason for World Hunger*. London: Penguin, 1986, p. 121.

[10] *Ibid*, p.196.

[11]"Public Health and Environmentalism: Adding Garbage to the History of Environmental Ethics," Christopher Preston and Steven Corey, *Environmental Ethics*, Spring, 2005, Vol. 27, p. 19.

[12] Allen Carlson, *Aesthetics and the Environment: The Appreciation of Nature, Art and Architecture*, New York: Routledge, 2000, p. 72.

[13] Malcolm Budd, "The Aesthetics of Nature," *Proceedings of the Aristotelian Society,* Vol. 100, No.2 2002. P.73.

[14] Plato uses the Greek word *anthropos* to indicate mankind as gender neutral rather than *andres* which would refer only to the male sex.

[15] Joshua Haddock, "*The Great Misguided Wilderness Debate: A Goal Oriented Approach to Environmental Policy,*" an unpublished paper presented at the Association for Practical and Professional Ethics, Feb. 2005, p.11.

[16] *Ibid* p.2.

[17] Light, Andrew and Katz, Eric, *Environmental Pragmatism*, New York, Routledge Publishing, 1996, p.1.

[18] Walton, Douglas, *The New Dialectic: Conversational Contexts of Arguments*, Toronto, University of Toronto Press, 1998, p. 151.

[19] The term teleology itself is derived from the Greek words *telos* and *logos* that mean the study or meaning of goals and carry within the terminology itself the connotation and denotation of goal oriented behaviors or explanations.

[20] Utopianism as a conceptual ideal is derived from two Greek words that mean a "good place" or the common translation that states it as "no where." Also, the term *aretean* within the context of norms is derived from the Greek word, *arete* meaning excellence that is the highest

standard for human aspiration and achievement.

[21] Kristen Carey, *Responsible Parenting and Reproductive Choice: A Stewardship Approach to Procreative Beneficence*, Villanova University, 2005.

[22] Statistically, the average IQ of a family decreases with an increase in quantity of the children. It is not clear, however, what percentage of this inverse correlation is caused by genetics alone.

[23] Phenotypes are essentially the results of the interaction between genetic inheritance and the environment in which they live.

[24] Brent Waters, "Engineering Our Grace: An Old Idea and New Genetic Technologies," in *Genetics and Ethics: An Interdisciplinary Study*, ed. Gerard Magill, St. Louis, MO, St. Louis University Press, 2004.

[25] Waters, *Reproductive Technology,* p. 49.

[26] Thomas Shepard, *Catalog of Teratogenic Agents,* 5th ed., Baltimore and London: John Hopkins University Press, 1986, p. xiii.

[27] Waters, "Engineering Our Grace," p. 126.

[28] It is instructive that the term stewardship is derived from the Anglo Saxon words, sty (pig) and caretaker. This humble etymology is earthy and does not suggest any ethereal origins. Also, the biblical reference to stewardship is found in St. Paul's letter to the Corinthians 6:14 wherein a willing preacher of the gospel is given a recompense whereas, an unwilling preacher of the gospel

is given a "stewardship " to clean out the pig sties as punishment.

[29] Waters, *Reproductive Technologies*, pp. 116-118.

[30] Julian Savalescu, "Procreative Beneficence: Why we should select the best children," *Bioethics* 15, no. 516, 2001, p. 415.

[31] Throughout the 19th century it was common in most countries of the world not to regard education as a basic need for survival and it was not part of the government's responsibilities to its citizens. In the United States, 95% of the population in 1900 were either living on subsistence farms or were one generation removed from farming, whereas in 2000 only 5% had that agrarian background.

[32] The long-standing controversy regarding the net energy needed to produce ethanol in contrast to petroleum fuel is addressed by Michael Wang's research at the Argonne National Laboratory published in 2004. It provided convincing evidence that ethanol yielded a net 25% energy gain over fossil fuels per net output due to the "free" solar energy photosynthesis of ethanol precursor grains such as corn, sugar cane or beets.

[33] "Can Technology Save the Planet?" Bruce Sterling, *Sierra*, July-August, Volume 90, No. 4, 2005, p. 32.

[34] "The Ghosts of Yosemite," Michelle Nuhuis, *High Country News*, October 17, 2005, p. 11.

[35] *Ibid*. p. 14.

[36] *Ibid*. p. 15.

[37] *Ibid.* p. 6.

[38] *World Commission on Environment and Development, Our Common Future*, Oxford, U.K.: Oxford University Press, 1987, p. 20.

[39] The law as a combination of legislation and litigation is ineffective and essentially unsuited to bring about ethics or morality-based environmental change. Introduction of morals and ethical teachings into legal forums overtly or covertly distorts the neutral principles of constitutional interpretation. Litigation represents the interests of plaintiffs and defendants within an adversarial arena while legislation is the result of the clearinghouse activities of special interest groups that are enforced by the coercive power of the sovereign. The sovereign is interested in keeping the peace, collecting taxes, recording legal instruments and maintaining its powers. Whether in a spectrum of cases pertaining to enforcement of the "takings clause" of the U.S. Constitution, gay marriages or environmental pollution issues, the law contemplates the recording and the filing of those contractual activities without pronouncing on their morality since it is not within its purview to do so. A citizen has no standing as a moral agent *per se* based on an alleged violation of personal ethical beliefs and cannot compete fairly on any level legislative field without the massive influence peddling power or the well financed PACs (political action committees) and special interest groups. In a word, morality or ethical teachings cannot be legislated or litigated without significant distortions of the legal process or

ethics.

[40] Charles Wechsler in his law review article, published in 1964, advances this ideal by calling for continued vigilance in upholding "neutral principles of constitutional interpretation" as a countervailing force to pernicious forms of self-serving influence, special interests and the distortions of naked political and financial power. However, fidelity to neutral principles of constitutional interpretation may not necessarily preclude certain inalienable rights (values) set forth in the Declaration of Independence and the Articles of Confederation. Specific value distinctions such as *mala in se* and *mala prohibita* that have a long-standing tradition and far reaching applications may also be contemplated within the context of neutral principles.

[41] Michael Reynolds, "Rendering Unto God," *Mother Jones*, Volume 30, No. 7, Dec., 2005. p. 79.

[42] *Ibid.*, p. 30.

[43] In *McCreary County v. the American Civil Liberties Union of Kentucky*, the central issue decided by the US Supreme Court involved a Ten Commandments display on the grounds of a county courthouse. The U.S. Supreme Court ordered this display to be dismantled based on the separation of church and state required by the U.S. Constitution.

[44] Karl Rahner, "Gefahrens," quoted in *The Thomist*, Vol. xxvii, 1963, The Thomist Press, Washington, D.C.

[45] Stanley Milgram's conclusions recounted in his 1964

book *Obedience to Authority* are still "shocking" in that randomly selected volunteers obediently increased the putative electrical voltage thought to be applied to a subject located behind a one-way glass partition. The authority figures, clad in white laboratory jackets were perceived by the volunteers to justify the departure from normal standards of behavior thereby giving permission to administer "near fatal" and excruciatingly painful “electrical shocks” to the "patients."

[46] *Code of Justinian* 1, viii, leg. 2 quoted in Edward Gibbon, *The Rise and Fall of the Roman Empire*, John Morris and Co., Philadelphia, PA, Vol. 3, p. 38.

[47] Aristocracy as the "rule of stars" has double meaning or *entendre deaux* as self-annointed "stars" in the social order and also due to the cosmic "star crossing" that bestowed special status by the constellations.

[48] "The Policy Turn in Environmental Philosophy," Robert Frodeman, *Environmental Ethics*, Spring, 2006. Vol. 28, No. 1, p. 10.

[49] The collapse of fish populations worldwide including Chesapeake Bay, the Grand Banks, Oregon Coast and the Gulf of Mexico has been the subject of the Pew Oceans Commission and the U.S. Commission on Ocean Policy in the 21st century. Over-fishing and global pollution, both examples of the "tragedy of the commons," are urgent and important issues in the United Nations Convention on the Law of the Sea document resolution that the U.S. Congress refused to pass in 2006.

[50] The political uses of science and technology are analyzed in *Frontiers of Illusion: Science, Technology and the Ideology of Progress* by Daniel Sarewitz, Philadelphia, Temple University Press, 1995.

[51] Washington, D. C., President's Council on *Bioethics*, 2003.

[52] The NSF's second review criteria can be found at http://www.bioethics.gov/reports/beyondtherapy.

[53] "Normal science" is a term of art that was introduced into discussions regarding science to explain ways in which scientists use paradigms agreed upon to define reality and to label what is real.

[54] Thomas Kuhn, *The Structure of Scientific Revolutions*, 2nd ed., 1970 Cambridge University Press, p. 77. Also, Karl Popper in his works regarding the scientific method has brought attention to the requirement of building a method of falsifying propositions used in scientific research. Popper's early application of falsifiability to Darwin's theory of natural selection led to his controversial conclusion that natural selection was an all-inclusive explanation that by explaining everything without being falsified, explained nothing.

[55] During outbreaks of the plague in England during the 17th century, tobacco smoking was recommended by the College of Physicians and mandated by the sovereign to abate this dread disease. Since that time, there have been a plethora of studies funded by tobacco companies that have consistently concluded that smoking tobacco is not

harmful to the smoker's health nor has negative health consequence for those breathing secondary smoke.

[56] Thomas Aquinas, a spokesperson for a traditional natural law tradition states the most general formulation of the natural law is, "Do good and avoid evil." From this broad ethical premise is derived additional prescriptive ethical directives, all of which are based by derivation on our knowledge of the natural world.

[57] "Exxon Mobil CEO Protects Profits by Trashing Good Science," Paul Krugman, *Salt Lake Tribune*, April 18, 2006. The extent of Exxon Mobil's profits is evidenced by its $8.4 billion profit realized in the 1st quarter of 2006.

[58] Mathew, Josephson,, *The Robber Barons, The Great American Capitalists,* 1861-1901, Harcourt, Brace, 1934, New York City, N.Y.

[59] App. 174, Cal. Rptr. 348, p.360 et. seq.

[60] *Grimshaw v. Ford Motor Co., op. cit.*, p. 361.

[61] Peter, Drucker, "What is Business Ethics?" *The Public Interest*, No. 64, Spring, 1981 p. 18.

[62] Kurt Baier, *The Moral Point of View: A Rational Basis of Ethics*, Random House, N.Y., N.Y. 1969, focuses on the details of this orientation and shows the interaction among moral perspectives, situational ethics, principles of action and the community of those taking a moral point of view guided by reason.

[63] William Frankena, *Talking About Morality*, Ann Arbor, University of Michigan Press, 1980, p. 26. Immanual

Kant noted that he was struck by the wonderment of the "starry skies above and the moral law within." He thought this awareness was a fundamental part of all shared human experience that, along with space and time were necessary as ways of knowing and understanding moral obligations.

[64] The word "religion," for example, is derived from the Latin *religare* that means to bind again. Citizens are initially bound at birth by the secular laws and are bound again when inducted into a religious system of rites, rituals and regulations.

[65] John Rawls' *Theory of Justice* is a laudable attempt to reintroduce venerable concepts such as justice, fairness and equitable considerations into ethical discussions. His 'theory' offers rational alternatives to greed with its gross and toxic inequities uncorrected by the 'invisible hand' or unrestrained by the hand of government. The elements of a contract model are evident as a basis for the minimum and maximum limits proposed by Rawls to acquisition and the excessive use of natural and social resources by certain individuals or groups.

[66] John Kenneth Gailbraith's *Affluent Society* written in 1959 focused on a central theme relevant to government intervention in behalf of the citizens. The dominance of corporations that produce, market and advertise consumer products creates demands for 'consumables' to the exclusion of commensurate demands for commonwealth needs including the many elements of a healthy environment. According to Gailbraith, government regulations and programs are necessary to

correct this imbalance.

[67] "The Ethanol Myth," *Consumer Reports*, Oct. 2006, p. 16.

[68] Mohammad Hadi & Karen Talley, "Socially Responsible Is in the Eye of the Investor," *Wall Street Journal*, 7-9-2006.

[69] *Salt Lake Tribune*, "Supreme Court Decision," November 30, 2006.

[70] The "bubble policy" of the Environmental Protection Agency so called because the bubbles of toxic gas freely floated into the environment as a business trade-off with less costly "bubble" controls. *Time*, 17 December 1979 p.71.

[71] The impact of this Executive Order is detailed in V. Kerry Smith, ed., *Environmental Policy under Reagan's Executive Order*, Chapel Hill, N.C. The University of North Carolina Press, 1984. Within this context of executive powers subverting legislation, the "signing statements" of the President are invoked. George W. Bush has executed over 800 such signing statements during his terms in office wherein the President does not veto legislation but states it will not be enforced.

[72] A representative and respected spokesperson for an "absolute ban ethics" during the timeframe of the Clean Air Acts is William Blackstone, "Equality and Human Rights," *Monist*, Vol. 52, No. 4, 1968. Blackstone's article and others generally rework Immanuel Kant's categorical imperatives and are based on unconditional human rights

that are unyielding to compromise or a nuanced public policy approach. In November, 2006, The United States Supreme Court began hearing arguments pertaining to global warming and pollution issues pursuant to the Clean Air Acts of 1967 and 1970. In the case, Massachusetts v. EPA, there are two central questions. 1) Do the Clean Air Acts authorize the EPA to regulate airborne pollutants that contribute to climate change. 2) If so, whether the EPA misused its authority in its refusal to enforce the provisions of the Clean Air Acts. After 35 years of non-enforcement 5 out of 9 Justices will decide a crucial issue of the 21st century.

[73] *The Hidden Epidemic: Sexually Transmitted Diseases*, published by the National Academy of Science, The National Institute of Health, Washington, D.C. in 1998 provides comprehensive documentation of these issues.

[74] Most all animal species show evidence of population control that manifests itself in various ways. Other than starvation that reduces populations after the fact, there is a general pattern of decreased conception, pregnancy and birth within species in direct relationship to the availability of food and water. An extreme adaptive population reduction strategy at certain population densities is noted in the mass destruction of lemmings that jump off cliffs into the sea. Despite rare examples of human population reduction due to shortage of food and water such as studies of the Irish potato famine 1845-51, human reproduction is generally not reduced by scarcity of essential resources. Thomas Malthus was not the first

political economist to observe this dysfunctional and potentially disastrous trait of humans. However, he articulated it with such erudition that it provided Charles Darwin the basic foundation for "natural selection" that emerges from the struggle for survival in a world of scarcity.

[75] Hannah Arendt's seminal book *The Human Condition* published in 1958 is best known for the delineation of the banality of evil within the context of the holocaust in Europe. The monstrous behaviors involved with these systematic agendas that obviated all minimal and/or traditional standards of civility and decency were attributed to the grossest forms of denial, self-deception, delusion and evil perpetrated in small increments over time. Habituation of these repetitious yet deviant actions made them banal yet had a cumulative effect that was repugnant to a social conscience and contrary to traditional and long-standing norms of moral good. In the year of her death in 2003, Arendt wrote an essay "Home to Roost" wherein she showed that associating the ideals of technological progress with an irresponsible consumer culture is at the "expense of the world in which we live…[in which]we no longer use but abuse, misuse and throw away." Hannah Arendt, "Home to Roost" in *Responsibility and Judgement,* New York, N.Y., Schoecken Books, 2003, p. 262.

Appendix

Appendix A

Primary Teratogenic Agents in Human Beings

Radiation
Therapeutic
Radioiodine
Atomic Weapons
Infections
Rubella virus
Cytomeyal virus
Herpes viruses I & II
Toxoplasmosis
Encephalitis virus
Syphillis
Maternal metabolic
Imbalance
Endemic creatinism
Diabetes
Phynylketonuria
Alcoholism
Hyperthermia
Rheumatic disease
Drugs and Environmental
Chemicals

Appendix A Continued

The grim reality of the teratogenic environmental agents is illustrated in this photograph of a child born deformed and blind from the effects of dioxin/agent orange. This defoliant was widely used during the Vietnam War for military purposes and has also been linked to many types of cancer, diabetes and neurological diseases. The euphemism of "collateral damage" from industrial chemicals, heavy metals and other toxins provides a verbal smoke-screen to obscure the significant issues of environmental ethics.

Appendix B

The middle fork of the Idaho Salmon River only had tourists floating the River in the 1940's. In the "wildest place" in the continental United States, visitors experience untrammeled nature, while scientists try to study, defend and preserve it.*

The waters of the Middle Fork are pristine because the watershed is protected from development. It is still vulnerable from contamination upstream as surrounding area becomes increasingly populated.

* An article "Captive Wilderness," Discover August, 2006 concluded that the ideal "wilderness concepts" of untrammeled nature is most nearly found in this geographical area.

Appendix C

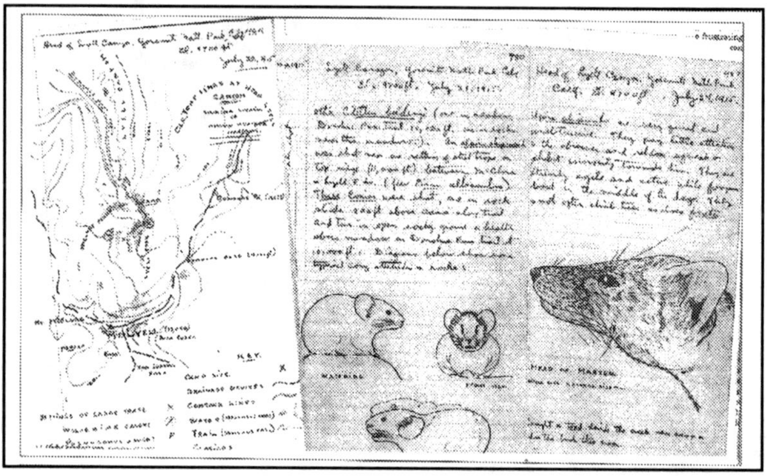

The detailed journal notes and drawings by wildlife biologist Charles Camp in 1915 provide a baseline for small animal populations in the Yosemite eco-system. Less than a century later these species and their habitat ranges have significantly deteriorated.

Appendix D

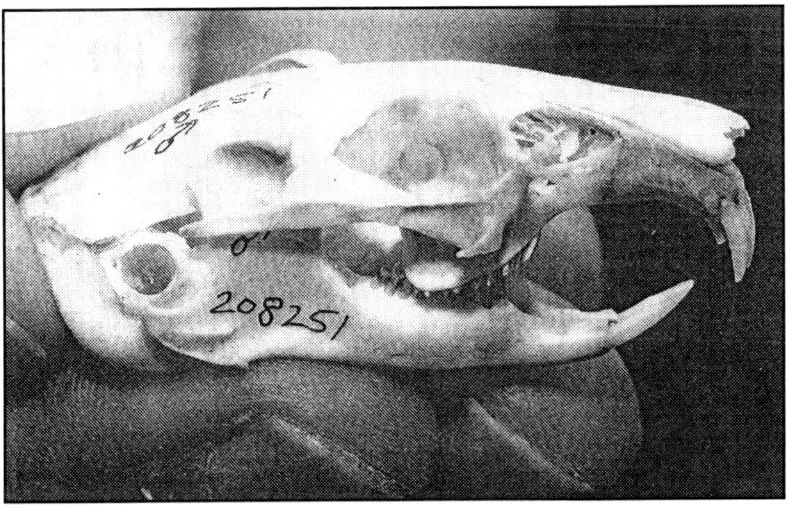

A typical skull of a small mammal marked by a field biologist is an example of increasing wildlife demise from toxic green house gases and global warming. Birds and other small animals are especially susceptible to climate change, toxins and polluted air. Their high metabolism rate makes them living sentinels to signal adverse conditions as did canaries carried into mine shafts.

Appendix E

Lyell Glacier in July, 1932 (above) has entirely melted by July 1988 (below). This is due to global warming especially pronounced over the past 30 years.

Appendix F

The Salt Lake Tribune OPINION Sunday, February 19, 2006 AA3

The LDS Church, extremism and the leadership vacuum

UTAH VOICES

During the waning months of Eisenhower's presidency, I was in Washington, D.C., working for Sen. Frank Church during the day, pursuing a graduate degree at night and campaigning for JFK against Nixon on Saturdays.

Sundays, I presided over the Sunday school in the LDS Washington Ward, attended by Secretary of Agriculture (and LDS Apostle) Ezra Taft Benson and his family.

JEROLD WILLMORE

America was locked in the Cold War. Worse, the fires of nuclear fear were fanned by McCarthyist "red-baiters," none more rabid than Robert Welch, president of the John Birch Society. Welch saw a communist behind every tree. He even called President Eisenhower "a communist dupe."

Benson's son, Reed, was the Washington representative for the Birch Society and a frequent speaker at Sacrament meetings, although his talks often reflected extremist right-wing political ideology.

During one such talk, he invited everyone to "an important conference" at the Marriott Hotel. About 40 of us showed up. Reed ran a 30-minute Birch Society recruiting film — a jingoistic diatribe by Welch himself.

Then Reed enthusiastically invited our questions but got an awkward silence. So, I rose to say that I found Welch's presentation demagogic, that Reed's super-patriotism was making an Americanist political front out of the LDS Church and that I wanted the record to show that he didn't speak for me. The meeting erupted in applause and broke up.

The Salt Lake Tribune ran the AP story the next morning. To its credit the LDS Church First Presidency responded with a statement strongly disapproving of its members supporting extremist organizations, whether of the right or the left.

The worldview of extremists is toxic because it's distorted by juvenile, two-valued thinking. They see everything simplistically, as black or white, good or evil, "you're either with us or the terrorists," etc. Real world subtlety, nuance and complexity are lost on extremists. For them the end justifies the means — even if it divides our country against itself.

Right-wing extremist fixation has shifted from anti-Communism to anti-terrorism. But their divisive tactics are unchanged. Their professed ends are still national security and cultural purity — and their means are still chauvinism and sanctimony. But if, as Samuel Johnson said, "Patriotism is the last refuge of a scoundrel," then surely, moralizing is the first. For history teaches that those who wrap themselves in the flag ("Mission Accomplished!") and parade their religiosity (Terry Schiavo?) are neither patriotic nor spiritual. They're sitting in judgment on the "moral decline" of everyone else.

The self-serving claims of the self-righteous to "moral clarity" and "super-patriotism" are the tactic; neoconservative extremism is the result. Unfortunately, neo(new)conservatism and the New Testament have nothing in common.

Would Jesus do, even reluctantly, what the neocons do zealously — enrich the rich and impoverish the poor, bury our grandchildren under public debt and violate the law to lie, spy, torture and commit unprovoked aggression? Hardly.

The neocons have prostituted the press, distorted science, compromised government and religion by entangling one with the other and stained America's reputation for moral leadership.

History teaches that those who wrap themselves in the flag ("Mission Accomplished!") and parade their religiosity (Terry Schiavo?) are neither patriotic nor spiritual. They're sitting in judgment on the "moral decline" of everyone else.

Worse, neocons persist in raping God's green Earth instead of protecting it. Result? Catastrophic global warming will soon be irreversible. The true enemies of humankind are neocons who would drill or war for more oil — instead of ending its consumption altogether. Neocon denial, while pretending to be pro-life, is extremist, because pro-life begins with pro-environment.

Utah's congressional delegation is a virtual rubber stamp for this faith-based, counterfeit Bush administration. Might this be a good time for the LDS Church to reassert its leadership against extremism? Here in the reddest state of Utah, that leadership could repair the unconstitutional gerrymandering that has emasculated Utah's and the nation's checks and balances, return us to civil moderation and heal the religious divide — and the Earth.

Jerold Willmore is president of a leadership training company in Salt Lake City

Appendix G

The terms *a priori* and *a posteriori* in this context refer to methodologies that are used to arrive at propositions fundamental to a system of thought. Generally, *a priori* is derived from Latin terms meaning before experience whereas *a posteriori* means based on experience. These experiences are the life experiences of persons that are constructed by the workings of the mind from sense data.

While there is an ancient history of these methodologies in a multiplicity of intellectual disciplines, there are common denominators that identify their applications. *A priori* methodologies yield inalienable rights articulated in constitutions, immutable ethical norms derived from essences and unqualified wilderness concepts.

A posteriori methodologies produce an array of contingent propositions that are built up by induction from many instances occurring over time. This methodology generates propositions subject to modification and translates to changing court decisions, situation or relative ethical views and malleable or flexible land use policies. Former U.S. Supreme Court Chief Justice, Oliver Wendell Holmes summarized the legal implications of these premises by stating that, "the life of the law is not logic but experience."

Appendix H

A graphic example of wolf species depopulation under sovereign indifference and the flauting of international wildlife preservation treaties is found in *Canis simensis*, an Ethiopian wolf species. The species evolved from European wolves that migrated to Africa during the Ice Age 100,000 years ago.

Appendix H
Continued

At the beginning of the 21st century, habitat destruction and diminishing food supplies along with outright killing of them have reduced a once thriving species to less than 600 wolves.

Appendix I

Because of the glaring inaction of sovereign and nation states leaders regarding environmental issues and public policy implementation, there is some evidence that "sub-sovereign" initiatives are being undertaken. The following text from a national conference summarizes some of these efforts.

The Salt Lake Tribune

Article Last Updated: 3/04/2006 01:27 AM

States' leadership cited in climate change concerns

U conference: Speaker says pressure on the federal government is coming from outside the Beltway

By Joe Baird
The Salt Lake Tribune

Salt Lake Tribune

Contrary to popular perception, the United States is beginning to rise to the challenge posed by climate change.

The catch is, the plans and policies to combat global warming are being created in statehouses and corporate board rooms - at least until the Bush administration and the Congress finally decide to jump into the game. And state and business actions will ultimately force them to do just that.

That was one of the messages conveyed Friday at the Stegner Center's Global Climate Change conference held at the S.J. Quinney College of Law on the campus of the University of Utah.

In the absence of any real leadership from Washington, governors and state legislatures have jumped in to fill the vacuum, Salt Lake City attorney James Holtkamp told a conference gathering of academics, lawyers, scientists and environmentalists.

"There are few places on the planet where the federal government does . . . what the public would like it to do, and the sub-sovereign states step up to do it. In every state, to at least some extent, climate change is now being addressed," said Holtkamp, who manages the environmental compliance group for the firm of Holland & Hart.

The acknowledged leader of the movement: California, which has been several steps ahead of everyone else for at least a decade. Now, with Republican Gov. Arnold Schwarzenegger calling for a series of reductions in greenhouse gas emissions over the next five decades - and an accompanying directive to auto manufacturers to cut carbon dioxide emissions in a similar manner - the Golden State is poised once again to be hugely influential.

"This is very important," Holtkamp said, "and it's a possible bellwether of what ends up happening nationally."

The ripples, he noted, are already being felt. States as disparate as New Mexico, Oregon, Massachusetts and Illinois are implementing emissions-reduction programs. Nine Northeast states have created a regional initiative to curb greenhouse gases. Arizona and New Mexico are partnering here in the West. And still other states are ramping up with action plans, though Utah has not yet joined the game.

The reduction programs can take many forms, including the trading of emissions credits, tax incentives and even leveraging market muscle to force out-of-state energy producers to deliver clean power - another recent emissions policy implemented by California.

Appendix I Continued

Sensing big changes on the horizon, Holtkamp says U.S. business is also beginning to rise to the global-warming challenge. In some cases, companies are attempting to position themselves for what's to come. Others are being prodded by their major shareholders. And still others are finding a more environmentally friendly approach to be more profitable.

In some ways, such a broad-based approach will prove beneficial, because ultimately there is no one solution for slowing down climate change.

"We will need many creative new policy proposals, because there's no silver bullet. No single approach will work everywhere in the world," said David Sandalow, director of the Brookings Institution's Environment and Energy Project.

Noting that there is just one current piece of federal legislation tied to global warming - an emissions cap bill proposed by Democrat Sen. Joseph Lieberman and Republican Sen. John McCain - Sandalow, too, senses that forward progress on climate change will come from further down the food chain.

"It's interesting that the governors of our two largest states, both Republican [Schwarzenegger and George Pataki of New York], are taking the most proactive approach to global warming. That tells you a lot about the politics of the issue," he said.

But Sandalow says federal leadership may not be far off, noting the Senate's passage of a nonbinding resolution calling for addressing the climate change problem. It may simply be a matter of waiting out the Bush administration - the major roadblock to federal global warming initiatives.

"There's a broadening of support for this issue," he said.

The two-day conference concludes today.

jbaird@sltrib.com

Appendix J

The impact of global warming and habitat loss is dramatized by the recent migration of the grizzly bear into polar bear range. Migration into the Far North arctic territory is 700 miles farther into Canada's Northwest Territories than previously recorded by wildlife biologists.

Appendix K

Benefits:

Savings	180 burn deaths, 180 serious burn injuries, 2100 burned vehicles.
Unit cost	$200,000 per death, $67,000 per injury, $700 per vehicle.
Total	180 x ($200,000) plus
Benefits	180 x ($ 67,000) plus
	2100 x ($ 700) = $49.15 million

Costs:

Sales	11 million cars, 1.5 million light trucks
Unit Cost	$11 per car, $11 per truck
Total	11,000,000 x ($11) plus
Costs	1,500,000 x ($11) = $137 million

From memorandum attached to statement of J.C. Echold, Director Ford Safety Division, Ford Motor Corporation[1]

Ford's estimate of the number of deaths, injuries, and vehicles that would be lost as a result of fires from fuel leakage were based on statistical studies. The $200,000 value attributed to the loss of life was based on a study of the National Highway Traffic Safety Administration, which broke down the estimated social costs of a death as follows:[2]

Component	1971 Costs
Future Productivity Losses	
Direct	$132,000
Indirect	41,300
Medical Costs	
Hospital	700
Other	425
Property Damage	1,500
Insurance Administration	4,700
Legal and Court	3,000
Employer Losses	1,000
Victim's Pain and Suffering	10,000
Funeral	900
Assets (Lost Consumption)	5,000
Miscellaneous Accident Cost	200
Total per Fatality	$200,725

[1] Ralph Drayton. "One Manufacturer's Approach to Automobile Safety Standards," *CTLA News*, VIII, no. 2 (February 1968): 11.
[2] Mark Dowie, "Pinto Madness," *Mother Jones* (September/October 1977), p. 28.

Appendix L

A nearly $100,000 gas guzzling SUV appears to obtrude itself into the domain of a beach etching made with a drift-wood stick. Can an environmental ethic become a social meme strong enough to prevent such an invasion into "off road" areas? Does the potential harm to persons, animals and places create categorical imperatives for a moral agent to rage against the machines, to sabotage them or stop them?

If a wilderness concept advocate were the skier in this pristine snow-scape would there be an ethical departure and a moral lapse based on the *aretean* wilderness ideal. Should the wilderness ethical idealists prevent other humans from making transient etchings or graffiti on the snow? Does it matter that non-renewable petroleum is used to produce the plastic boots and other synthetic polymers or metal mechanisms used by the skier?

Appendix L
Continued

The Wilderness Acts of 1964 included language that prohibits the use of mechanized or mechanical devices in the wilderness areas. Recently, there have been challenges brought by "mixed use" advocates urging the use of mountain bikes, dirt bikes and some motorized vehicles such as ATV's, motorcycles and snowmobiles despite the language of the Acts. Mechanical vehicle advocates argue that cross-country skiers and those using snowshoes employ mechanical devices in their bindings and this would constitute unfair (invidious) discrimination against the mechanized and motorized advocates.

Appendix M

A credible environmental ethic includes provisions for bio-diversity as a wilderness concept and as nuanced public policy. Diverse mammalian forms of life require minimum environmental quantity and quality to make a living and to sustain their lives and life styles.

Transportation energy to transport tourists to these "crystal clear waters" is nearly 100% fossil fuel worldwide which fuel produces massive concentrations of carbons, particulate matter, greenhouse gases and global warming. Some countries are producing electricity almost entirely without fossil fuel consumption. France, for example is 80% nuclear and 20% hydroelectric; Canada is 50%-50%; Sweden is 50% hydroelectric and 40% nuclear; Denmark is 18% windturbine generated electrical power and 82% coal.

Appendix N

THE CURVED FIN *of an orca (opposite) surfing beside a whale-watching boat off Northern Vancouver Island, Canada, indicates a female. Dorsal fins on adult males are straight and up to 6 feet long. Whale watchers get their money's worth as an orca breaches off Canada's Stewart Island (below). A decline in chinook salmon, a key prey species, jeopardizes the survival of orcas in Washington State's Puget Sound.*

Multiple pollutants, global warming, shipping traffic and declining food sources have combined to sharply decrease Orcas populations in many of their habitats. Heavy metal contamination and increased water temperatures have yielded a 50% mortality rate for salmon, the primary food for Orcas. Both the salmon and Orca have some of the highest levels of carcinogenic polychorinated biphenyls (PCBs) and polybrominated diphenyl ethers of any animals in the world. The author states, "We ignore this looming environmental problem at our own peril. The Orcas are the ultimate indicator of the health of the marine eco-system."

Appendix O

"Wildebeests of the Serangeti," Richard Estes Natural History, Volume 115, No. 7 9/06.

Newborn wildebeest gains its feet just a few minutes after birth. Even at the height of the calving season, the large aggregation of animals must keep moving from pasture to pasture. Hence, unlike the offspring of most antelopes, the young cannot adopt the strategy of remaining concealed from predators. Instead, they must be able to move almost immediately with the herd.

Appendix O
Continued

A controlled population is sustained by various animal species including the African wildebeest *(tauriaus mearusi)* and the Arctic caribou evolving on separate continents. Over 80% of all calves of these species are born annually within a 2-3 week period. For the caribou, calving season is an adaptation to the very short Arctic summer that increases calf survival. The wildebeest over-produces calves during this short calving period, thereby glutting the predators and also controlling their populations by this evolutionary strategy. Despite these adaptive methods, the survival of these species and others depends on a sustainable equilibrium that requires sufficient water, forage and migration habitation all of which are being depleted.

BIBLIOGRAPHY

BOOKS

Arendt, Hannah, *The Human Condition,* Chicago, IL, University of Chicago Press, 1958.

Baier, Kurt, *The Moral Point of View: A Rational Basis of Ethics*, New York, N.Y., Random House, 1969.

Beyond Therapy: Biotechnology and the Pursuit of Happiness, Washington, D.C., President's Council on Bioethics, 2003.

Beauchamp, Tom and Childress, James, *Principles of Biomedical Ethics*, 2nd ed., New York, N.Y., Oxford University Press, 1989, 1994, 2001.

Bishop, William, *International Law: Cases & Materials,* 2nd ed. Boston & Toronto, Little Brown and Co., 1962.

Callicott, Baird and Nelson, Michael, *The Great Wilderness Debate*, Athens, GA., University of Georgia Press, 1998.

Carlson, Allen, *Aesthetics and the Environment: The Appreciation of Nature, Art and Architecture*, New York: Routledge, 2000.

Carson, Rachel, *Silent Spring*, Urbana, IL; Mariner Books, 2002.

Daly, Herman, *Toward a Steady State Economy,* San Francisco, CA, W.H. Freeman, 1973.

Darwin, Charles, *On Origins of Species*, London, Blackwell, 1968.

Erlich, Paul, *The Population Bomb*, Palo Alto; Stanford University Press, 1960.

Fastinger, Leon, *A Theory of Cognitive Dissonance*, Evanston, IL, White Plains, N.Y. Rove, Peterson and Co., 1957.

Frankena, William, *Talking About Morality*, Ann Arbor, MI., University of Michigan Press, 1980.

Gailbraith, John Kenneth, *The Affluent Society*, Cambridge, M.A., Harvard University Press, 1959.

George, Susan, *How the Other Half Dies: The Real Reason for World Hunger*, London: Penguin, 1986.

Gibbon, Edward, *The Rise and Fall of the Roman Empire*, Philadelphia, PA, John Morris and Co., Vol.3.

Hagman, Donald, *Public Planning and Control of Urban and Land Development:* Cases and Materials. St. Paul,

MN., West Publishing Co. 1973.

Hursthouse, Rosalind, *On Virture Ethics*, N.Y. Oxford University Press, 1999.

Johnson, Phillip, *Darwin on Trial*, Downers Grove, IL, Inter-varsity Press, 1991.

Jones, W.T., ed, *A History of Western Philosophy*, New York, N.Y. Hardcourt, Brace and World Inc., 1969

Jowett, Benjamin, *Dialogues of Plato*, Volumes 1-3, London, Oxford Press, 1982

Josephson, Mathew, *The Robber Barons, The Great American Capitalists*, 1861-1901, New York, N.Y., Harcourt, Brace, 1934.

Light, Andrew and Katz, Eric, *Environmental Pragmatism*, New York, N.Y. Routledge Publishing, 1996.

Malcolm I, Thomas, *The Luddities: Machine Breaking in Regency England*, New York, N.Y.; Schocken, 1972.

Marx, Karl, *Early Writings*, New York, N.Y. McGraw Hill, 1964.

Kuhn, Thomas, *The Structure of Scientific Revolutions*, 2nd ed., Cambridge, M.A. Cambridge University Press,

1970.

Rawls, John, *Theory of Jusice*, Cambridge, M.A., Harvard University Press, 1973.

Santler, Ronald and Cafaro, Phillip ed., *Environmental Virtue Ethics*, Lantham, M.D. Rowman and Littlefield, 2005.

Sarewitz, Daniel, *Frontiers of Illusion: Science, Technology and the Ideology of Progress*, Philadelphia, PA., Temple University Press, 1995.

Shepard, Thomas, *Catalog of Teratogenic Agents*, 5th ed., Baltimore and London, John Hopkins University Press, 1986.

Smith,V.Kerry, *Environmental Policy Under Reagan's Executive Order*, Chapel Hill, N.C., The University of North Carolina Press, 1984.

Swanton, Christine, *Virtue Ethics: A Pluralistic View*, N.Y., Oxford University Press, 2003.

Velasquez, Manuel, *Business Ethics: Concepts and Cases*, 2nd ed., Princeton, N.J., Prentice Hall, 1988.

Walton, Douglas, *The New Dialectic: Conversational Contexts of Arguments*, Toronto, University of Toronto Press, 1998.

Waters, Brent, *Engineering our Grace: An Old Idea and New Genetic Technologies in Genetics and Ethics: An Interdisciplinary Study*, ed. Gerard Marill, St. Louis, MO, St. Louis University Press, 2004.

Wittgenstein, Ludwig, *Philosophical Investigations*, Oxford: Blackwell Press, 1968.

Articles and Authorities

Blackstone, William, *Equality and Human Rights,* Monist, Vol. 52, No. 4, 1968.

Budd, Malcolm, *The Aesthetics of Nature*, Proceedings of the Aristotelian Society, 100, no. 2, 2002.

"Bubble Policy", Environmental Protection Agency, Time, December, 17, 1979, Business Week, July 9, 1984.

The Ethanol Myth, Consumer Report, Oct. 2006.

Carey, Kristen, *Responsible Parenting and Reproductive Choice: A Stewardship Approach to Procreative Beneficence*, APPE 14th Annual Colloquium, Feb., 2005.

Corey, Steven, and Preston, Christopher, *Public Health and Environmentalism: Adding Garbage to the History of Environmental Ethics,* Environmental Ethics, Vol. 27, No. 4, Spring 2005.

Drucker, Peter, *What is Business Ethics?*, The Public Interest, No. 64, Spring 1981.

Duncan, David Ewing, *The Pollution Within*, National Geographics, Vol. 210, No. 4, Oct., 2006.

Frodeman, Robert, *The Policy Turn in Environmental*

Philosophy, Environmental Ethics, Vol. 28, No.1, Spring 2006.

Haddock, Joshua, *The Great Misguided Wilderness Debate: A Goal Oriented Approach to Environmental Policy,* Association for Practical and Professional Ethics, Feb., 2005.

Hardin, Garrett, *Living on a Lifeboat*, cited in Environmental Ethics Vol. 26, No. 4, 2004.

Hardin, Garrett, *The Tragedy of the Commons* in Herman Daly, ed., *Toward a Steady State Economy*, San Francisco: W.H. Freeman, 1973.

James, Simon, *Human Virtues and Natural Values*, Environmental Ethics, Winter 2006, Vol. 28, No. 4.

Mohammad Hadi & Karen Talley, *'Socially Responsible' Is in the Eye of the Investor*, Wall Street Journal, July 9, 2006.

Nolt, John, *The Move from Good to Ought*, in Environmental Ethics, Winter 2006, Vol. 28, No. 4.

Nuhuis, Michelle, *The Ghosts of Yosemite*, High Country News, p.11, October 17, 2005.

Olsen, Kew, *Orcas on the Edge*, National Wildlife, Vol., 44, No. 6., Oct/Nov., 2006.

Peterson, Anna, *Toward a Materialist Environmental Ethics*, Environmental Ethics, Winter, 2006. Vol. 28, No. 4.

Rahner, Karl, *"Gefahrens,"* quoted in The Thomist, Vol. Xxvii; 1963, The Thomist Press, Washington, D.C.

Reynolds, Michael, *Rendering Unto God*, Mother Jones, Volume 30, No. 7, Dec. 2005.

Rodman, John, *The Liberation of Nature*, Inquiry 20, 1977.

Singer, Peter, *Famine, Affluence and Morality,* Philosophy and Public Affairs, J. No. 3, 1972.

Savalescu, Julian, *Procreative Beneficence: Why we should select the best children*, Bioethics 15, no.516, 2001.

Sterling, Bruce, *Can Technology Save the Planet?*, Sierra, Volume 90, No. 4, July-August, 2005.

INDEX

A

B

C

D

E

F

G

J

K

L

M

P

Q

R

U

V

W

Richard Crangle is the founder and CEO of Crantec Research, a research and development company for innovative products. He is an adjunct professor at the University of Utah, Salt Lake Community College and Westminster College, and has been a faculty member at Augustana College and Creighton University. In addition to giving workshops, Richard is a consultant for several entities, inclusing the University of Utah and private corporations.

Through Crantec Research, Richard has invented, researched and developed many proprietary medical technologies, and has been a co-inventor of several innovative technology devices. His experience includes inventing, developing, producing, marketing and istributing products over a 15-year period. In the area of technology innovation, Richard is a veteran and has experienced success and loss, both of which contribute to a balanced treament of the technology innovation business and the patent process. He has written five books in his Invention Works Series available in major book-stores and through Amazon.com and Barnes & Noble.com.

Crantec Research
1366 Murray-Holladay Road
Salt Lake City, Utah 84117

email: rncrangle@protectnewtech.com
web: www.razorpages.com

Printed in the United States
93623LV00002B/472-519/A

9 780966 483550